Lecture Notes in Computer Science 5892

Commenced Publication in 1973
Founding and Former Series Editors:
Gerhard Goos, Juris Hartmanis, and Jan van Leeuwen

Editorial Board

Krzysztof Janowicz   Martin Raubal
Sergei Levashkin (Eds.)

GeoSpatial Semantics

Third International Conference, GeoS 2009
Mexico City, Mexico, December 3-4, 2009
Proceedings

 Springer

Volume Editors

Krzysztof Janowicz
The Pennsylvania State University, Department of Geography
University Park, PA 16802, USA
E-mail: k@janowicz.de

Martin Raubal
University of California, Department of Geography
Santa Barbara, CA 93106, USA
E-mail: raubal@geog.ucsb.edu

Sergei Levashkin
National Polytechnic Institute, Centro de Investigacion en Computacion
07738 Mexico City, Mexico
E-mail: sergei@cic.ipn.mx

Library of Congress Control Number: 2009939281

CR Subject Classification (1998): J.2, J.4, H.2.8, H.3, H.2, H.4, I.2.9

LNCS Sublibrary: SL 3 – Information Systems and Application, incl. Internet/Web and HCI

ISSN 0302-9743

ISBN 978-3-642-10435-0 Springer Berlin Heidelberg New York

springer.com

© Springer-Verlag Berlin Heidelberg 2009

Typesetting: Camera-ready by author, data conversion by Scientific Publishing Services, Chennai, India
Printed on acid-free paper   SPIN: 12800599   06/3180   5 4 3 2 1 0

Preface

GeoS 2009 was the third edition of the International Conference on Geospatial Semantics. It was held in Mexico City, December 3-4, 2009.

Within the last years, geospatial semantics has become a prominent research field in GIScience and related disciplines. It aims at exploring strategies, computational methods, and tools to support semantic interoperability, geographic information retrieval, and usability. Research on geospatial semantics is a multidisciplinary and heterogeneous field, which combines approaches from the geosciences with philosophy, linguistics, cognitive science, mathematics, and computer science. With the increasing popularity of the Semantic Web and especially the advent of linked data, the need for semantic enablement of geospatial services becomes even more pressing. In general, semantic interoperability plays a role if data are acquired in a different context than they are finally used for. This is the case when shifting from the document Web to the data Web. The core idea of linked data is to make information contributed by various actors, with different cultural backgrounds, and different applications in mind available to the public. Understanding, matching, and translating between the conceptualizations underlying these data becomes a key challenge for future research on geospatial semantics.

This volume contains full research papers, which were selected from among 19 submissions received in response to the Call for Papers. Each submission was reviewed by three or four Program Committee members and 10 papers were chosen for presentation. The papers focus on foundations of geo-semantics, the formal representation of geospatial data, semantics-based information retrieval and recommender systems, spatial query processing, as well as geo-ontologies and applications. Overall, a diverse body of research was presented coming from institutions in Austria, Germany, Mexico, The Netherlands, Spain, Taiwan, and the USA.

We are in debt to many people who made this event happen. The members of the Program Committee offered their help with reviewing submissions. Our thanks go also to Miguel Matinez, Nahun Montoya, Walter Renteria, Iyeliz Reyes, and Linaloe Sarmiento who formed the Local Organizing Committee and took care of all the logistics. The Centro de Investigación en Computación, Mexico City, Mexico, was the local host and co-sponsored GeoS 2009. Finally, we would like to thank all the authors who submitted papers to GeoS 2009, Christoph Stasch, Arne Bröring, and Pascal Hitzler for giving tutorials about Sensor Web Enablement and rules in OWL, respectively, as well as our keynote speakers Andrew Frank and Pascal Hitzler.

December 2009
Krzysztof Janowicz
Martin Raubal
Sergei Levashkin

Organization

Organizing Committee

Conference Chair

Sergei Levashkin
Centro de Investigación en Computación
Mexico City, Mexico

Program Chairs

Krzysztof Janowicz
The Pennsylvania State University, USA
Martin Raubal
University of California, Santa Barbara, USA

Organizing Chairs

Centro de Investigación en Computación
Mexico City, Mexico
Miguel Matinez (Chair)
Nahun Montoya
Walter Renteria
Iyeliz Reyes
Linaloe Sarmiento

GeoS 2009 Program Committee

Neeharika Adabala	Microsoft Research India, India
Ola Ahlqvist	The Ohio State University, USA
Naveen Ashish	University of California, Irvine, USA
Brandon Bennett	University of Leeds, UK
Ioan Marius Bilasco	Laboratoire d'Informatique Fondamentale de Lille, France
Stefano Borgo	National Research Council, Italy
Boyan Brodaric	Geological Survey of Canada, Canada
Gilberto Camara	INPE, Brazil
Isabel Cruz	University of Illinois at Chicago, USA
Clodoveu Davis	Universidade Federal de Minas Gerais, Brazil
Andrew Frank	Technical University Vienna, Austria
Mark Gahegan	The University of Auckland, New Zealand
Brent Hecht	Northwestern University, Chicago, USA
Cory Henson	Wright State University, USA
Stephen Hirtle	University of Pittsburgh, USA
Pascal Hitzler	Wright State University, USA
Prateek Jain	Wright State University, USA
Tomi Kauppinen	Helsinki University of Technology, Finland

Marinos Kavouras	National Technical University of Athens, Greece
Carsten Keßler	University of Münster, Germany
Alexander Klippel	The Pennsylvania State University, USA
Craig Knoblock	University of Southern California, USA
Margarita Kokla	National Technical University of Athens, Greece
Dave Kolas	BBN Technologies, USA
Werner Kuhn	University of Münster, Germany
Felix Mata	Centro de Investigación en Computación Mexico City, Mexico
Marco Painho	ISEGI Universidade Nova de Lisboa, Portugal
Christine Parent	École publique Polytechnique Fédérale de Lausanne, Switzerland
Vasily Popovich	St. Petersburg Institute for Informatics and Automation of the Russian Academy of Sciences, Russia
Sudha Ram	University of Arizona, USA
Christoph Schlieder	University of Bamberg, Germany
Angela Schwering	University of Münster, Germany
Shasi Shekhar	University of Minnesota, Germany
Kathleen Stewart Hornsby	The University of Iowa, USA
Nancy Wiegand	University of Wisconsin-Madison, USA
Stephan Winter	The University of Melbourne, Australia

Sponsoring Institutions

Instituto Politécnico Nacional (IPN), Mexico
Centro de Investigación en Computación (CIC), Mexico
National Council on Science and Technology (CONACYT), Mexico

Table of Contents

Geo-ontologies and Applications

Multi-cultural Aspects of Spatial Knowledge

Andrew U. Frank

Geoiformation TU Wien, Gusshausstrasse 27-29/E127
frank@geoinfo.tuwien.ac.at

It is trivial to observe differences between cultures: people use different languages, have different modes of building houses and organize their cities differently, to mention only a few. Differences in the culture of different people were and still are one of the main reasons for travel to foreign countries. The question whether cultural differences are relevant for the construction of Geographic Information Systems is longstanding (Burrough et al. 1995) and is of increasing interest since geographic information is widely accessible using the web and users volunteer information to be included in the system (Goodchild 2007). The review of how the question of cultural differences was posed at different times reveals a great deal about the conceptualization of GIS at different times and makes a critical review interesting.

At the heart of the discussion of cultural differences relevant for GIScience is a Whorfian hypothesis that different cultural backgrounds could be responsible for differences in the way space and spatial relations are conceived. Whorf claimed that people using a language with more differentiation, for example in terms describing different types of snow, also perceive reality differently from people using a language with less differentiation (Carroll 1956). An early contribution picked up on suggestions made by Mark and others (1989b) and identified several distinct issues that could be investigated individually (Campari et al. 1993):

1. the cultural assumptions that are built into the GIS software may differ from those of the user;
2. the influence of decision context in which a GIS is used;
3. the conceptualization of space and time may differ;
4. differences in the administrative processes and how they structure space;
5. the sense of territoriality, ownership or dominance of space, is different between people, again citing ethnographic examples;
6. the influence of the material culture, the ecosystem, economy and technology.

Campari and Frank in this early paper asked the question whether a single or a few GIS software packages could serve universally or local (national) development of GIS software, which still existed at that time, were justified by cultural differences.

1 Initial Focus on Cognitive Cultural Differences

Montello (1995) concentrated on cultural differences in the conceptualization of space and argued that a large share of spatial cognition is universal, i.e., the same for all human beings, because the problems the environment posed to humans during their

K. Janowicz, M. Raubal, and S. Levashkin (Eds.): GeoS 2009, LNCS 5892, pp. 1–8, 2009.

development and to which their cognitive apparatus adapted, is basically the same for all humans; Montello refers to substantial empirical evidence for this claim. Evidence to the contrary had, despite efforts, not be found. For example the study by Freundschuh investigated whether growing up in a regular "Manhattan" grid would influence the spatial cognition compared with other teenagers who grew in a modern, curved road suburban setting; the results were not conclusive (Freundschuh 1991).

Linguistics has explored the different ways that the languages of the world express spatial relations. Well known are the central—periphery organizations used in Hawaii (Mark et al. 1989b), the use of up-down in valleys or on slopes (Bloom et al. 1994). Montello also addresses the Whorfian hypothesis and mentions the lack of evidence that 'language structures space' in a direct way (as a paper title by Talmy (1983) may be misunderstood to suggest). The differences in methods to express spatial situations in different languages, e.g., the preference for egocentric or absolute frames (Levinson 1996; Frank 1998a; Klatzky 1998), are observed in situations where no best solution exists and are preferences rather than absolute choices: Western cultures prefer an egocentric expression (the glass on the left) whereas others, often rural groups prefer cardinal directions (Perderson (1993) for India, observe also the use of cardinal directions in a fight between two men in Synge's play 'The Playboy of the Western World' in Anglo-Irish); these are only *preferences* for one method and the other method is available as well. Montello's contribution suggests for the GIS development that multiple software modules that recognize and work with spatial situations could be useful universally and no cultural adaptation for differences in spatial conceptualization is likely necessary.

Egenhofer gave mathematical definitions for spatial terminology to express topological relations between regions and to make precise studies of what natural language terms like 'touch' mean possible (Egenhofer 1989; Egenhofer et al. 1991). He defined, for example, a large number of differentiable topological relations between a region and a line. With Mark he observed how people would group these relations in groups that are differentiated in verbal expressions (Mark et al. 1992; Mark et al. 1994). The testing situation asked questions about a road (the line) and a park (the region); it must be suspected that the context created in the testing situation, affects the grouping—separating cases that contain differences that are practically relevant in the situation. More tests could be worthwhile to see how context influences, but tests to discover cultural differences were not successful (Mark et al. 1995a) and revealed more commonality (Ragni et al. 2007).

Comparable to the formalization of topological relations are efforts to construct qualitative distance and direction relations (Frank 1992; Freksa 1992; Zimmermann 1993; Hernández et al. 1995; Zimmermann et al. 1996) from which qualitative spatial reasoning emerged as a subfield of spatial information theory. This line of research produced typically tables showing the result of composing two (or more) relations: e.g., Santa Barbara is west of Los Angeles and Los Angeles is west of New York, therefore we can conclude that Santa Barbara is west of New York.

The research in spatial cognition applied to GIS was driven by a hope that natural language like communication with humans would become the way we interact with a GIS; the influential paper 'Naive Geography' by Mark and Egenhofer discussed the differences between formal and human conceptualizations (Mark et al. 1995a). It was expected that computer programs could correct for typical human incorrect conceptualizations (e.g., alignment error (Stevens et al. 1978) in the coast line around Santa Barbara, which runs conceptually north-south, but geographically east-west). Despite well-documented, regularly observed cases, where human and formal definitions systematically differ, no formalization has been published so far, but it was found, that these typical human spatial reasoning errors are independent of cultures. For example, Xiao found similar effects of regionalization in China as is observed in western cultures (Xiao et al. 2007).

2 Linguistic Differences

Users of GIS with a native language different than the one used for the user interface of the software could encounter difficulties and errors and misunderstandings result. Campari investigated (1994) the command language of GIS and the differences that result when non-native speakers are confronted with command language terms originating from English. It showed in detail that a native Italian speaker could misunderstand the meaning of translated command language terms because the connotations and metaphors evoked are different. For example the English term 'layer' has different connotations than the corresponding Italian term 'copertura'. The concern in the early 1990 was that spatial professionals would have only a basic knowledge of English and would use translated manuals and command language; today, GIS specialists learn the English based GIS command as they learn other computer terms, fully aware of the limits of metaphorical transfer of common-sense knowledge to the virtual realm. The difficulty Campari pointed to is absorbed by the trained GIS specialist who builds the application for users and bridges the differences from the English based technical GIS vocabulary to the user's description of operations in his language.

The differences between the vocabulary appear simpler: Different vocabulary terms seem to describe the same class of things and translation a simple mapping: from French 'chien' to German 'Hund' to English 'dog'. Unfortunately, for most terms, translation is not as simple: the English use the two terms 'in' and 'on' whereas German differentiates 'in' 'an' and 'auf' (Frank 1998b) and a direct mapping fails: Germans ride 'in' the train or bus, whereas English ride 'on' the train or bus. In a landmark paper Mark (1993) compares natural language terms for landscape features in English, French, and Spanish; it becomes apparent that these closely related languages use different distinctions (Frank 2006) to comparable (but not strictly translatable) landscape terms. Despite the clear-cut definitions in dictionaries, comparison with the pragmatics of landscape terms, i.e., their actual use, is strongly influenced by the ecological context. His work was based on dictionary definitions, but comparing actual use of such terms in toponymes casts doubts on the strictness of the definitions (Mark personal communication).

Current research of Mark and his colleagues in ethnophysiography investigates landscape terminology used by indigenous people in different parts of the world (Mark et al. 2007). They are careful to select people in similar ecological situations (arid regions in south western USA and in northwest of Australia) to reduce the effects of ecotopes. An observation surprising us is that a stream-bed and the water flowing in it can be strongly separated conceptually. They also observe a strong tendency to 'populate' the landscape with spirits (ghosts), which is a reflection of a polytheistic religion and thus a cultural difference.

3   Differences in the Spatial Structure and the Physical Environment

GIS still force our understanding of the world to fixed, exactly bounded objects. This is, on one hand, the effect of the use of coordinated geometry, and on the other, the inheritance from a land use planning tradition, where land use is planned for non-overlapping, clearly bounded regions. Legal traditions differ in how sharp they create boundaries; current European law varies between a concept of general boundary that is formed by a (possible wide) hedge in England and Wales and geometrically sharp boundary lines fixed by coordinates in Austria.

A variety of aspects were discussed in a workshop (Burrough et al. 1996b); resulting in reports that show the counter-intended effects sharp boundaries can have (Burrough et al. 1996a). Campari (1996) discussed the conflict between clear-cut two dimensional planning regions as they are applicable in the physical environment for which the earliest planning GIS were built in the 1970s, i.e., the U.S. Midwest suburban towns and their planning, and applications of GIS to capture the reality of traditional towns, built on steep inclines, e.g., in southern Europe. The limited two dimensional view is insufficient and a three dimensional representation is necessary, but not likely resulting in sharply delimited and single use regions that can be entered in a GIS; the 'open' space in a town serves for transportation, access to sunlight but also for rainwater runoff. The application to GIS in other cultures, with other building styles, climates etc. may require other deviations from the two dimensional, sharp boundary model.

Efforts to define approximate spatial relations between vague regions are important to bridge the gap between the geometric reasoning of GIS and the human users. Sharma described in his PhD thesis an approximate calculus for distance and directions (1996) following the approach by Frank (1992) and Rezayan et al. (2005).

Specialized systems for spatial navigation in cars but increasingly also for pedestrians have become very popular and questions of how humans give directions are now practically important (Lovelace et al. 1999) studies in the 1980s (Denis 1997) have shown (small) differences between genders, but no differences between, say European and USA (unpublished thesis TU Wien). State of the art in commercial devices give satisfactory wayfinding instructions in simple cases, but they are not satisfactory in complex situations where the spatial reasoning by the system differs widely from the way humans conceptualize a situation. To give precise verbal instructions to navigate a complex, multi-bifurcation is assisted by using graphical displays—distracting the driver in a situation where his attention should be on the other moving cars around him; equally distracting are the differences between the system's view of where a turn instruction is necessary and where not—indicating that the concept of 'following a road' based on the road classification and numbering scheme and the visual perceived reality conflict.

It is apparent that geographic information could be used to improve the search on the web. In many cases, a query has a spatial focus and objects satisfying the conditions, but far away, are not relevant (e.g., search for a pizza place, or an ATM). To process queries like 'show me the pizza places downtown' or 'find a hotel in the black forest' we require a definition of where 'downtown' (Montello et al. 2003) or what the 'black forest' is. Efforts to glean this information statically from the use of such terms on the web are underway and often reported as using 'vernacular' location terms (to differentiate from the toponymes collected in official gazetteers) (Twaroch et al. 2008).

The context dependence of qualitative spatial relations is well known but poorly understood. Most of the above issues to make GIS more usable and more 'user friendly' depend on understanding how the present context influences the meaning of the terms used. Linguists have studied context dependence of semantics in general and have—unfortunately—not come up with a satisfactory answer. A recent publication by Gabora, Rosch and Aerts (2008) gives a very precise account of the difficulties of previously proposed approaches and sketches a novel method; it uses quantum mechanics as a calculus for transforming expressions between different contexts and claims that it corresponds to empirical observations. The application to spatial situations is a promising, but open question.

4 Conclusions

Differences between cultures affect how GIS is used and 'cultural differences' form a major obstacle in the application of GIS. The initial fears of substantial differences in the cognition of space by human beings from different cultures has not been confirmed. Similarly, the way spatial relations are described appears more situation (context) dependent than culturally different. Studies formalizing spatial reasoning showed very substantial "cultural" differences between the way a computer system treats geometry and human performance, a field of research, were many questions are still open—but fortunately simplified in so far as we may expect that human spatial cognition and human spatial reasoning is universal. Differences in the conceptualization of spatial situations—independent from the socio-economic (cultural) context— are not documented, but large differences in language expressions to communicate spatial situations exist. A possible explanation is that the conceptualization and the mental classification is much finer and only for communication mapped to the coarser verbal expression. Early research in spatial cognition in GIS assumed a close connection between the mental concepts and the verbal expressions (Mark et al. 1989a) and followed a linguistic tradition to reported the verbal expression as spatial concept.

Large differences exist in the way, spatial information is used in different cultures. The practice of spatial decision making is different, because the cultural (social, legal, economic) situation provides a different context and requires different distinctions (Frank 2006) between objects to form classes of situations that can be dealt with similarly. Such cultural differences are visible between countries—especially those using the same language—but are also observable between different agencies in a single city, or even between different parts of a single organization. To understand multi-cultural influences in geographic information, research today could be focused on the following aspects:

- Differences in the vocabulary: terminology differs (lake vs. lac, in the distinctions used to separate the terms, e.g., small-large to separate pond and lake, vs. l'etang separated from lac by man made—natural;
- Differences in graphical style, perhaps most obvious in the cartographic styles found in maps of different National Mapping Agencies. Some of the differences follow from the landscape, ecological, economical and political situation;
- Cultural meaning of terms; evident are the differences even between countries sharing the same language (USA, Canada, India, UK, etc. or Germany, Austria and (part of) Switzerland). The cultural (legal) environment defines concepts and terms, which are meaningful in this social-cultural context (X counts as Y in the context Z—(Searle 1995)) and differ widely, even when using the same word.
- Differences in conversation style (Grice 1989): is it acceptable to anthropomorphizing computers? Levels of politeness are required even in a computer dialog. Length of turns between the partners in a conversation.

References

Bloom, P., Peterson, M.A., Nadel, L., Garrett, M.F. (eds.): Language and Space. Language, Speech and Communication. MIT Press, Cambridge (1994)

Burrough, P., Couclelis, H.: Practical Consequences of Distinguishing Crisp Geographic Objects. In: Masser, I., Salgé, F. (eds.) Geographic Objects with Indeterminate Boundaries, pp. 333–335. Taylor & Francis, London (1996a)

Burrough, P.A., Frank, A.U.: Concepts and Paradigms in Spatial Information: Are Current Geographic Information Systems Truly Generic? International Journal of Geographical Information Systems 9(2), 101–116 (1995)

Burrough, P.A., Frank, A.U. (eds.): Geographic Objects with Indeterminate Boundaries. GIS-DATA Series. Taylor & Francis, London (1996b)

Campari, I.: GIS Commands as Small Scale Space Terms: Cross-Cultural Conflicts of Their Spatial Content. In: SDH 1994, Sixth International Symposium on Spatial Data Handling, Association for Geographic Information, Edinburgh, Scotland (1994)

Campari, I.: Uncertain Boundaries in Urban Space. In: Burrough, P.A., Frank, A.U. (eds.) Geographic Objects with Indeterminate Boundaries, vol. 2, pp. 57–69. Taylor & Francis, London (1996)

Campari, I., Frank, A.U.: Cultural differences in GIS: A basic approach. In: EGIS 1993, Genoa, March 29 - April 1, EGIS Foundation (1993)

Carroll, J.B.: Language, Thought and Reality - Selected Writing of Benjamin Lee Whorf. The MIT Press, Cambridge (1956)

Denis, M.: The Description of Routes: A cognitive approach to the production of spatial discourse. Cahiers de psychologie cognitive 16(4), 409–458 (1997)

Egenhofer, M.J.: Spatial Query Languages. PhD University of Maine (1989)

Egenhofer, M.J., Franzosa, R.D.: Point-Set Topological Spatial Relations. International Journal of Geographical Information Systems 5(2), 161–174 (1991)

Frank, A.U.: Qualitative Spatial Reasoning about Distances and Directions in Geographic Space. Journal of Visual Languages and Computing (3), 343–371 (1992)

Frank, A.U.: Formal models for cognition - taxonomy of spatial location description and frames of reference. In: Freksa, C., Habel, C., Wender, K.F. (eds.) Spatial Cognition 1998. LNCS (LNAI), vol. 1404, pp. 293–312. Springer, Heidelberg (1998a)

Frank, A.U.: Specifications for Interoperability: Formalizing Spatial Relations 'In', 'Auf' and 'An' and the Corresponding Image Schemata 'Container', 'Surface' and 'Link'. In: Proceedings of 1st Agile-Conference, ITC, Enschede, The Netherlands (1998b)

Frank, A.U.: Distinctions Produce a Taxonomic Lattice: Are These the Units of Mentalese? In: International Conference on Formal Ontology in Information Systems (FOIS), Baltimore, Maryland. IOS Press, Amsterdam (2006)

Freksa, C.: Using Orientation Information for Qualitative Spatial Reasoning. In: Frank, A.U., Formentini, U., Campari, I. (eds.) GIS 1992. LNCS, vol. 639, pp. 162–178. Springer, Heidelberg (1992)

Freundschuh, S.M.: Spatial Knowledge Acquisition of Urban Environments from Maps and Navigation Experience. PhD Buffalo (1991)

Gabora, L., Rosch, E., Aerts, D.: Toward an Ecological Theory of Concepts. Ecological Psychology 20(1), 84–116 (2008)

Goodchild, M.: Citizens as Sensors: the World of Volunteered Geography. GeoJournal 69(4), 211–221 (2007)

Grice, P.: Logic and Conversation. In: Studies in the Way of Words, pp. 22–40. Harvard University Press, Cambridge (1989)

Hernández, D., Clementini, E., Di Felice, P.: Qualitative Distances. In: Kuhn, W., Frank, A.U. (eds.) COSIT 1995. LNCS, vol. 988, pp. 45–57. Springer, Heidelberg (1995)

Klatzky, R.L.: Allocentric and egocentric spatial representations: definitions, distinctions, and interconnections. In: Freksa, C., Habel, C., Wender, K.F. (eds.) Spatial Cognition 1998. LNCS (LNAI), vol. 1404, pp. 1–17. Springer, Heidelberg (1998)

Levinson, S.C.: Frames of Reference and Molyneux's Question: Crosslinguistic Evidence. In: Bloom, P., Peterson, M.A., Nadel, L., Garett, M.F. (eds.) Language and Space, pp. 109–170. MIT Press, Cambridge (1996)

Lovelace, K.L., Hegarty, M., Montello, D.R.: Elements of Good Route Directions in Familiar and Unfamiliar Environments. In: Freksa, C., Mark, D.M. (eds.) COSIT 1999. LNCS, vol. 1661, p. 65. Springer, Heidelberg (1999)

Mark, D.M.: Toward a Theoretical Framework for Geographic Entity Types. In: Campari, I., Frank, A.U. (eds.) COSIT 1993. LNCS, vol. 716, pp. 270–283. Springer, Heidelberg (1993)

Mark, D.M., Comas, D., Egenhofer, M., Freundschuh, S., Gould, M.D., Nunes, J.: Evaluating and Refining Computational Models of Spatial relations through Cross-Linguistic Human-Subjects Testing. In: Kuhn, W., Frank, A.U. (eds.) COSIT 1995. LNCS, vol. 988, pp. 553–568. Springer, Heidelberg (1995)

Mark, D.M., Comas, D., Egenhofer, M.J., Freundschuh, S.M., Gould, M.D., Nunes, J.: Evaluating and Refining Computational Models of Spatial Relations Through Cross-Linguistic Human-Subjects Testing. In: Kuhn, W., Frank, A.U. (eds.) COSIT 1995. LNCS, vol. 988, pp. 553–568. Springer, Heidelberg (1995b)

Mark, D.M., Egenhofer, M.J.: An Evaluation of the 9-Intersection for Region-Line Relations. In: GIS/LIS 1992 Proceedings, ACSM-ASPRS-URISA-AM/FM, San Jose (1992)

Mark, D.M., Egenhofer, M.J.: Calibrating the Meanings of Spatial Predicates from Natural Languages: Line-Region Relations. In: Sixth International Symposium on Spatial Data Handling, Edinburgh, Scotland (1994)

Mark, D.M., Frank, A.U.: Concepts of Space and Spatial Language. In: Auto-Carto 9, ASPRS & ACSM, Baltimore, MA (1989a)

Mark, D.M., Frank, A.U., Egenhofer, M.J., Freundschuh, S.M., McGranaghan, M., White, R.M.: Languages of Spatial Relations: Initiative Two Specialist Meeting Report. National Center for Geographic Information and Analysis (1989b)

Mark, D.M., Turk, A.G., Stea, D.: Progress on Yindjibarndi Ethno-Physiography. In: Winter, S., Duckham, M., Kulik, L., Kuipers, B. (eds.) COSIT 2007. LNCS, vol. 4736, pp. 1–19. Springer, Heidelberg (2007)

Montello, D.R.: How Significant are Cultural Differences in Spatial Cognition? In: Kuhn, W., Frank, A.U. (eds.) COSIT 1995. LNCS, vol. 988, pp. 485–500. Springer, Heidelberg (1995)

Montello, D.R., Goodchild, M., Gottsegen, J., Fohl, P.: Where's Downtown? Behavioral Methods for Determining Referents of Vague Spatial Queries. Spatial Cognition and Computation 3(2&3), 185–204 (2003)

Pederson, E.: Geographic and Manipulable Space in Two Tamil Linguistic Systems. In: Campari, I., Frank, A.U. (eds.) COSIT 1993. LNCS, vol. 716, pp. 294–311. Springer, Heidelberg (1993)

Ragni, M., Tseden, B., Knauff, M.: Cross-Cultural Similarities in Topological Reasoning. Springer, Heidelberg (2007)

Rezayan, H., Frank, A.U., Karimipour, F., Delavar, M.R.: Temporal Topological Relationships of Convex Spaces in Space Syntax Theory. In: International Symposium on Spatio-Temporal Modeling 2005, Beijing, China, Hong Kong Polytechnic University (2005)

Searle, J.: The Construction of Social Reality. The Free Press, New York (1995)

Sharma, J.: Integrated Spatial Reasoning in Geographic Information Systems. PhD Maine (1996)

Stevens, A., Coupe, P.: Distortions in judged spatial relations. Cognitive Psychology 10, 422–437 (1978)

Talmy, L.: How Language Structures Space. In: Pick, H., Acredolo, L. (eds.) Spatial Orientation: Theory, Research, and Application. Plenum Press, New York (1983)

Twaroch, F., Jones, C.B., Abdelmoty, A.I.: Acquisition of a vernacular gazetteer from web sources. In: Boll, S., Jones, C.B., Kansa, P., et al. (eds.) Proceedings of the First International Workshop on Location and the Web, LocWeb, vol. 300, pp. 61–64. ACM, New York (2008)

Xiao, D., Liu, Y.: Study of Cultural Impacts on Location Judgments in Eastern China. In: Winter, S., Duckham, M., Kulik, L., Kuipers, B. (eds.) COSIT 2007. LNCS, vol. 4736, pp. 20–31. Springer, Heidelberg (2007)

Zimmermann, K.: Enhancing Qualitative Spatial Reasoning - Combining Orientation and Distance. In: Campari, I., Frank, A.U. (eds.) COSIT 1993. LNCS, vol. 716, pp. 69–76. Springer, Heidelberg (1993)

Zimmermann, K., Freksa, C.: Qualitative Spatial Reasoning Using Orientation, Distance, and Path Knowledge. Applied Intelligence 6, 49–58 (1996)

Towards Reasoning Pragmatics

Pascal Hitzler

Kno.e.sis Center, Wright State University, Dayton, Ohio
http://www.pascal-hitzler.de/

Abstract. The realization of Semantic Web reasoning is central to substantiating the Semantic Web vision. However, current mainstream research on this topic faces serious challenges, which force us to question established lines of research and to rethink the underlying approaches.

1 What Is Semantic Web Reasoning?

The ability to combine data, mediated by metadata, in order to derive knowledge which is only implicitly present, is central to the Semantic Web idea. This process of accessing implicit knowledge is commonly called *reasoning*, and formal model-theoretic semantics tells us exactly what knowledge is implicit in the data.[1]

Let us attempt to define reasoning in rather general terms: *Reasoning is about arriving at the exact answer(s) to a given query.* Formulated in this generality, this encompasses many situations which would classically not be considered reasoning – but it will suffice for our purposes. Note that the definition implicitly assumes that there *is* an exact answer. In a reasoning context, such an exact answer would normally be defined by a model-theoretic semantics.[2]

Current approaches to Semantic Web reasoning, however, which are mainly based on calculi drawn from predicate logic proof theory, face several serious obstacles.

- Scalability of algorithms and systems has been improving drastically, but systems are still incapable of dealing with amounts of data on the order of magnitude as can be expected on the World Wide Web. This is aggrevated by the fact that classical proof theory does not readily allow for parallelization, and that the amount of data present on the web increases with a similar growth rate as the efficiency of hardware.
- Realistic data, in particular on the web, is generally noisy. Established proof-theoretic approaches (even those including uncertainty or probabilistic methods) are unable to cope with this kind of data in a manner which is ready for large-scale applications.

---

[1] It is rather peculiar that a considerable proportion of so-called Semantic Web research and publications ignores formal semantics. Even most textbooks fail to explain it properly. An exception is [7].

[2] Simply referring to a *formal* semantics is too vague, since this would also include procedural semantics, i.e. non-declarative approaches, and thus would include most mainstream programming languages.

K. Janowicz, M. Raubal, and S. Levashkin (Eds.): GeoS 2009, LNCS 5892, pp. 9–25, 2009.

– It is a huge engineering effort to create web data and ontologies which are of sufficiently high quality for current reasoning approaches, and usually beyond the abilities of application developers. The resulting knowledge bases are furthermore severely limited in terms of reusability for other application contexts.

The state of the art shows no indications that approaches based on logical proof theory would overcome these obstacles anytime soon in such a way that large-scale applications on the web can be realized. Since reasoning is central to the Semantic Web vision, we are forced to rethink our traditional methods, and should be prepared to tread new paths.

A key idea to this effect, voiced by several researchers (see e.g. [3,23]) is to explore alternative methods for reasoning. These may still be based more or less closely on proof-theoretic considerations, or they may not. They could, e.g., utilize methods from statistical machine learning or from nature-inspired computing.

Researchers who are used to thinking in classical proof-theoretic terms are likely to object to this thought, arguing that a relaxation of strict proof-theoretic requirements on algorithms, such as soundness and completeness, would pave the way for arbitrary algorithms which do not perform logical reasoning at all, and thus would fail to adhere to the specification provided by the formal semantics underlying the data – and thus jeopardize the Semantic Web vision. While such arguments have some virtue, it needs to be stressed that the nature of the underlying algorithm is, effectively, unimportant, as long as the system adheres to the specification, i.e. to the formal semantics.

Imagine, as a thought experiment, a black box system which performs sound and complete reasoning in all application settings it is made for – or at least up to the extent to which standard reasoning systems are sound and complete.[3] Does it matter then whether the underlying algorithm is *provably* sound and complete? I guess not. The only important thing is that its *performance* is sound and complete. If the black box were orders of magnitude faster then conventional reasoners, but somebody would tell you that it is based on statistical methods, which one would you choose to work with? Obviously, the answer depends on the application scenario – if you'd like to manage a bank account, you may want to stick with the proof-theoretic approach since you can prove formally that the algorithm does what it should; but if you use the algorithm for web search, the quicker algorithm might be the better choice. Also, your choice will likely depend on the evidence given as to the correctness of the black box algorithm in application settings.

This last thought is important: If a reasoning system is not based on proof theory, then there must be a quality measure for the system, i.e., the system must

---

[3] Usually, they are not sound and complete, although they are based on underlying algorithms which are, theoretically, sound and complete. Incompleteness comes from the fact that resources, including time, are limited. Unsoundness comes from bugs in the system.

be evaluated against the gold standard, which is given by the formal semantics, or equivalently by the provably sound and complete implementations [23].

If we bring noisy data, as on the web, into the picture, it becomes even clearer why a fixation on soundness and completeness of reasoning systems is counter-productive for the Semantic Web: In the presence of such data, even the formal model-theoretic semantics breaks down, and it is quite unclear how to develop algorithms based on proof theory for such data. The notions of soundness and completeness of reasoning in the classical sense appear to be almost meaningless. But only almost, since alternative reasoning systems which are able to cope with noisy data can still be evaluated against the gold standard on non-noisy data, for quality assurance.

In the following, we revisit the role of soundness and completeness for reasoning, and argue further for an alternative perspective on these issues (Section 2). We also discuss key challenges which need to be addressed in order to realise reasoning on and for the Semantic Web, in particular the questions of expressivity of ontology languages (Section 3), roads to bootstrapping (Section 4), knowledge acquisition (Section 5), and user interfacing (Section 6). We conclude in Section 7.

2   The Role of Soundness, Completeness, and Computational Complexity

Computational complexity has classically been a consideration for the development of description logics, which underlie the Web Ontology Language OWL – which is currently the most prominent ontology language for Semantic Web reasoning. In particular, OWL is a decidable logic. The currently ongoing revision OWL 2 [6] furthermore explicitly defines fragments, called profiles, with lower (in fact, polynomial) computational complexity.

Soundness and completeness are central properties of classical reasoning algorithms for logic-based knowledge representation languages, and are thus central notions in the development of Semantic Web reasoning around OWL. However performance issues have prompted researchers to advocate *approximate reasoning* for the Semantic Web (see e.g. [3,23]). Arguing for this approach provokes radically different kinds of reactions: some logicians appear to be abhorred by the mere thought, while many application developers find it the most natural thing to do. Often it turns out that the opposing factions misunderstand the arguments: counterarguments usually state that leaving the model-theoretic semantics behind would lead to arbitrariness and thus loss of quality. So let it be stated again explicitly: approximate reasoning shall not replace sound and complete reasoning in the sense that the latter would no longer be needed. Quite in contrast, approximate reasoning in fact needs the sound and complete approaches as a gold standard for evaluation and quality assurance. The following shall help to make this relationship clear.

2.1 Sound but Incomplete Reasoning

There appears to be not much argument against this in the Semantic Web com-mmunity, even from logicians: they are used to this, since some KR languages, including first-order predicate logic, are only semi-decidable,[4] i.e. completeness can only be achieved with unlimited time resources anyway. For decidable languages, however, a sound but incomplete reasoner should always be evaluated against the gold standard, i.e., against a sound and complete reasoner, in order to show the amount of incompleteness incurred versus the gain in efficiency. Interestingly, this is rarely done in a structured way, which is, in my opinion, a serious neglect. A statistical framework for evaluation against the gold standard is presented in [23].

2.2 Unsound but Complete Reasoning

Allowing for reasoning algorithms to be unsound appears to be much more controversial, and the usefulness of this concept seems to be harder to grasp. However, there are obvious examples. Consider, e.g., fault-detection in a power plant in case of an emergency: The system shall determine (quickly!) which parts of the factory need to be shut down. Obviously, it is of highest importantance that the critical part is contained in the shutdown, while it is less of a problem if too many other parts are shut down, too.[5] Another obvious example is semantic search: In most cases, users would prefer to get a quick set of replies, among which the correct one can be found, rather than wait longer for one exact answer.

Furthermore, sound-incomplete and unsound-complete systems can sometimes be teamed up and work in parallel to provide better overall performance (see e.g. [24]).

2.3 Unsound and Incomplete Reasoning

Following the above arguments to their logical conclusion, it should become clear why unsound and incomplete reasoning has its place among applications. Remember that there is the gold standard against which such systems should be evaluated. And obviously there is no reason to stray from a sound and complete approach if the knowledge base is small enough to allow for it.

The most prominent historic example for an unsound and incomplete yet very successful reasoning system is Prolog. Traditionally, the unification algorithm, which is part of the SLD-resolution proof procedure used in Prolog [13], is used without the so-called *occurs check*, which, depending on the exact implementation, can cause unsoundness [16].[6] This omission was made due to reasons of efficiency, and turned out to be feasible since it rarely causes a problem for Prolog programmers.

---

[4] Some non-montonic logics are not even semi-decidable.

[5] This example is due to Frank van Harmelen, personal communication.

[6] To obtain a wrong answer, execute the query ?-p(a). on the logic program consisting of the two clauses p(a) :- q(X,X). and q(X,f(X))., e.g. under SWI-Prolog. – The example is due to Markus Krötzsch.

Likewise, it is not unreasonable to expect that carefully engineered unsound and incomplete reasoning approaches can be useful on the Semantic Web, in particular when sound and complete systems fail to provide results within a reasonable time span. Furthermore, there is nothing wrong with using entirely alternative approaches to this kind of reasoning, e.g., approaches which are not based on proof theory.

To give an example of the latter, we refer to [2], where the authors use a statistical learning approach using support vector machines. They train their machine to infer class membership in $\mathcal{ALC}$, which is a description logic related to OWL, and achieve a 90% coverage. Note that this is done without any proof theory, other than to obtain the training examples. In effect, their system learns to reason with high coverage without performing logical deduction in the proof-theoretic sense.

For a statistical framework for evaluation against the gold standard we refer again to [23].

2.4   Computational Complexity and Decidability

Considerations on computational complexity and decidability have been driving research around description logics, which underlie OWL, from the beginning. At the same time, there are more and more critical voices concerning the fixation of that research on these issues, since it is not quite clear how practical systems would benefit from these. Indeed, theoretical (worst-case) computational complexity is hardly a relevant measure for the performance of real systems. At the same time, decidability is only guaranteed assuming bug-free implementations– which is an unrealistic assumption –, and given enough resources – which is also unrealistic since the underlying algorithms often require exponential time in the worst case.

The misconception underlying these objections is that computational complexity and decidability are not practial measures which have a direct meaning in application contexts. They are rather a priori measures for language and algorithm development, and the recent history of OWL language development indicates that these a priori measures have indeed done a good job. It is obviously better to have such theoretical means for the conceptual work in creating language features, than to have no measures at all. And indeed this has worked out well, since e.g. reasoning systems based on realistic OWL knowledge bases currently seem to behave rather well despite the high worst-case computational complexity.

Taking the perspective of approximate reasoning algorithms as laid out earlier, it is actually a decisively positive feature of Semantic Web knowledge representation languages that systems exist which can serve as a gold standard reference. Considering the difficulties in other disciplines (like Information Retrieval) in creating gold standards, we indeed are delivered the gold standard on a silver plate. We can use this to an advantage.

3   Diverse Knowledge Representation Issues

Within 50 years of KR research, many issues related to the representation of non-classical knowledge have been investigated. Many of the research results obtained in this realm are currently carried over to ontology languages around OWL, including abductive reasoning, uncertainty handling, inconsistency handling and paraconsistent reasoning, closed world reasoning and non-monotonicity, belief revision, etc.

However all these approaches face the same problems that OWL reasoning faces, foremost scalability and the dealing with realistic noisy data.[7] Indeed under most of these approaches, runtime performance becomes worse, since the reasoning problems generally become harder in terms of computational complexity.

Nevertheless, research on logical foundations of these knowledge representation issues, as currently being carried out, is needed to establish the gold standard. At this time there is a certain neglect, however, in combining several paradigms, e.g. it is quite unclear how to marry paraconsistent reasoning with uncertainty handling.

Research into enhancing expressivity of ontology languages can roughly be divided into the following.

- Classical logic features: This line of research follows the well-trodden path of extending e.g. OWL with further expressive features, while attempting to retain decidability and in some cases low computational complexity. Some concrete suggestions for next steps in this direction are given in the appendix.
- Extralogical features: These include datatypes and additional datastructures, like e.g. Description Graphs [19].
- Supraclassical logic: Logical features related to commonsense reasoning like abduction and explanations (e.g., [8]), paraconsistency (e.g., [15]), belief revision, closed-world (e.g., [4]), uncertainty handling (e.g., [10,14]), etc. There is hardly any work investigating approximate reasoning solutions for supraclassical logics.

Investigations into these issues should first establish the gold standard following sound logical principles including computational complexity issues. Only then should extensions towards approximate reasoning be done.

4   Bootstrapping Reasoning

How to get from A (today) to B (reasoning that works on the Semantic Web)? I believe that a promising approach lies in bootstrapping existing applications which use little or no reasoning, based e.g. on RDF. The idea is to enhance these applications very carefully with a bit more reasoning, in order to clearly

---

[7] I'm personally critical about fuzzy logic and probabilistic logic approaches in practice for Semantic Web issues. Dealing with noisy data on the web does not seem to easily fall in the fuzzy or probabilistic category. So probably new ideas are needed for these.

understand the added value and the difficulties one is facing when doing this. A (very) generic workflow for the bootstrapping may be as follows.

1. Identify an (RDF) application where enhanced reasoning would bring added value.
2. Identify ontology language constructs which would be needed for expressing the knowledge needed for the added value.
3. Identify an ontology language (an OWL profile or an OWL+Rules hybrid) which covers these additional language constructs.
4. Find a suitable reasoner for the enhanced language.
5. Enhance the knowledge base and plug the software components together.

The point of these exercises is not only to show that more reasoning capabilities bring added value, but also to identify obstacles in the bootstrapping process.

5 Overcoming the Ontology Acquisition Bottleneck

The ontology acquisition bottleneck for logically expressive ontologies comes partly from the fact that sound and complete reasoning algorithms work only on carefully devised ontologies, and in many cases it needs an ontology expert to develop them. Creating such high-quality ontologies is very costly.

A partial solution to this problem is related to (1) noise handling and (2) the bootstrapping idea. With the current fixation on sound and complete reasoning it cannot be expected that usable ontologies (in the sound and complete sense) will appear in large quantities e.g. on the web. However, it is conceivable that e.g. Linked Open Data[8] (LoD) could be augmented with more expressive schema data to allow e.g. for reasoning-based semantic search. The resulting extended LoD cloud would still be noisy and not readily usable with sound and complete approaches. So reasoning approaches which can handle noise are needed. This is also in line with the bootstrapping idea: We already have a lot of metadata available, and in order to proceed we need to make efforts to enhance this data, and to find robust reasoning techniques which can deal with this real-world noisy data.

6 Human Interfacing

Classical ontology engineering often has the appearance of expert sysem creation, if used off the web. On the web, it often lacks the reasoning component.

As argued in this paper, in order to advance the Semantic Web vision – on and off the web – we need to find ways to reason with noisy and incomplete data in a realistic and pragmatic manner. This necessity is not reflected by current ontology engineering research.

---

[8] http://esw.w3.org/topic/SweoIG/TaskForces/CommunityProjects/LinkingOpenData/

In order to advance towards ontology reasoning applications, we need ontology engineering systems which

- support reasoning bootstrapping,
- include multiple reasoning support (i.e. multiple reasoning algorithms, classical and non-classical), and
- are made to cope with noisy and uncertain data.

We need to get away from thinking about ontology creation as coding in the programming sense. This can only be achieved by relaxing the reasoning algorithm requirements, i.e. by realising reasoning systems which can cope with noisy and uncertain data.

7   Putting It All Together

In this paper I argue for roads to realising reasoning on and for the Semantic Web. Efforts on several fronts are put forth:

- The excellent research results and ongoing efforts in establishing sound and complete proof-theory-based reasoning systems need to be complemented by investigations into alternative reasoning approaches, which are not necessarily based on proof theory, and can handle noisy and uncertain data.
- Reasoning bootstrapping should be investigated seriously and on a broad front, in order to clearly show added value, and in order to identify challenges in adopting ontology reasoning in applications.
- Ontology engineering environments should systematically accommodate reasoning bootstrapping and the support of multiple reasoning paradigms, classical and non-classical.

Let us recall a main point of this paper: reasoning algorithms do not have to be based on proof theory. But they have to perform well if compared with the gold standard.

In a sense, the laid out lines of research lead us a bit further away from knowledge representation (KR), and at the same time they do a small step towards non-KR-based intelligent systems: Not all the intelligence must be in the knowledge base (with corresponding sound and complete reasoning algorithms). We must facilitate intelligent solutions, including machine learning and data mining, and statistical inductive methods, to achieve the Semantic Web vision.

Acknowledgement

Many thanks to Cory Henson, Prateek Jain, and Valentin Zacharias for feedback and discussions. The appendix is taken from [5].

References

1. Boley, H., Kifer, M. (eds.): RIF Framework for Logic Dialects. W3C Working Draft, July 30 (2008), http://www.w3.org/TR/rif-fld/
2. Fanizzi, N., d'Amato, C., Esposito, F.: Statistical learning for inductive query answering on OWL ontologies. In: Sheth, A.P., Staab, S., Dean, M., Paolucci, M., Maynard, D., Finin, T., Thirunarayan, K. (eds.) ISWC 2008. LNCS, vol. 5318, pp. 195–212. Springer, Heidelberg (2008)
3. Fensel, D., van Harmelen, F.: Unifying reasoning and search to web scale. IEEE Internet Computing 11(2), 96, 94–95 (2007)
4. Grimm, S., Hitzler, P.: A preferential tableaux calculus for circumscriptive ALCO. In: Proceedings of the Third International Conference on Web Reasoning and Rule Systems, Washington D.C., USA. LNCS. Springer, Heidelberg (to appear, 2009)
5. Hitzler, P.: Suggestions for OWL 3. In: Proceedings of OWL – Experiences and Directions, Sixth International Workshop, Washington D.C., USA (October 2009) (to appear)
6. Hitzler, P., Krötzsch, M., Parsia, B., Patel-Schneider, P.F., Rudolph, S. (eds.): OWL 2 Web Ontology Language: Primer. W3C Proposed Recommendation, September 22 (2009), http://www.w3.org/TR/2009/PR-owl2-primer-20090922/
7. Hitzler, P., Krötzsch, M., Rudolph, S.: Foundations of Semantic Web Technologies. Chapman & Hall/CRC (2009)
8. Horridge, M., Parsia, B., Sattler, U.: Laconic and precise justifications in OWL. In: Sheth, A.P., Staab, S., Dean, M., Paolucci, M., Maynard, D., Finin, T., Thirunarayan, K. (eds.) ISWC 2008. LNCS, vol. 5318, pp. 323–338. Springer, Heidelberg (2008)
9. Horrocks, I., Patel-Schneider, P.F., Boley, H., Tabet, S., Grosof, B., Dean, M.: SWRL: A Semantic Web Rule Language. W3C Member Submission, May 21 (2004), http://www.w3.org/Submission/SWRL/
10. Klinov, P., Parsia, B.: Optimization and evaluation of reasoning in probabilistic description logic: Towards a systematic approach. In: Sheth, A.P., Staab, S., Dean, M., Paolucci, M., Maynard, D., Finin, T., Thirunarayan, K. (eds.) ISWC 2008. LNCS, vol. 5318, pp. 213–228. Springer, Heidelberg (2008)
11. Krötzsch, M., Rudolph, S., Hitzler, P.: Description logic rules. In: Ghallab, M., Spyropoulos, C.D., Fakotakis, N., Avouris, N. (eds.) Proceedings of the 18th European Conference on Artificial Intelligence, ECAI 2008, pp. 80–84. IOS Press, Amsterdam (2008)
12. Krötzsch, M., Rudolph, S., Hitzler, P.: ELP: Tractable rules for OWL 2. In: Sheth, A.P., Staab, S., Dean, M., Paolucci, M., Maynard, D., Finin, T., Thirunarayan, K. (eds.) ISWC 2008. LNCS, vol. 5318, pp. 649–664. Springer, Heidelberg (2008)
13. Lloyd, J.W.: Foundations of Logic Programming. Springer, Heidelberg (1987)
14. Lukasiewicz, T., Straccia, U.: Managing uncertainty and vagueness in description logics for the semantic web. Journal on Web Semantics 6(4), 291–308 (2008)
15. Ma, Y., Hitzler, P.: Paraconsistent reasoning for OWL 2. In: Proceedings of the Third International Conference on Web Reasoning and Rule Systems, Washington D.C., USA, October 2009. LNCS. Springer, Heidelberg (to appear, 2009)
16. Marriott, K., Sondergaard, H.: On Prolog and the occur check problem. SIGPLAN Not. 24(5), 76–82 (1989)
17. McGuinness, D.L., van Harmelen, F. (eds.): OWL Web Ontology Language Overview. W3C Recommendation, February 10 (2004), http://www.w3.org/TR/owl-features/

18. Motik, B., Cuenca Grau, B., Horrocks, I., Wu, Z., Fokoue, A., Lutz, C. (eds.): OWL 2 Web Ontology Language: Profiles. W3C Proposed Recommendation, September 22 (2009), http://www.w3.org/TR/2009/PR-owl2-profiles-20090922/
19. Motik, B., Cuenca Grau, B., Horrocks, I., Sattler, U.: Representing Ontologies Using Description Logics, Description Graphs, and Rules. Artificial Intelligence 173(14), 1275–1309 (2009)
20. Motik, B., Patel-Schneider, P.F., Parsia, B. (eds.): OWL 2 Web Ontology Language: Structural Specification and Functional-Style Syntax. W3C Candidate Recommendation, September 22 (2009),
http://www.w3.org/TR/2009/PR-owl2-syntax-20090922/
21. Motik, B., Sattler, U., Studer, R.: Query-answering for OWL-DL with rules. Journal of Web Semantics 3(1), 41–60 (2005)
22. Rudolph, S., Krötzsch, M., Hitzler, P.: Cheap Boolean role constructors for description logics. In: Hölldobler, S., Lutz, C., Wansing, H. (eds.) JELIA 2008. LNCS (LNAI), vol. 5293, pp. 362–374. Springer, Heidelberg (2008)
23. Rudolph, S., Tserendorj, T., Hitzler, P.: What is approximate reasoning? In: Calvanese, D., Lausen, G. (eds.) RR 2008. LNCS, vol. 5341, pp. 150–164. Springer, Heidelberg (2008)
24. Tserendorj, T., Rudolph, S., Krötzsch, M., Hitzler, P.: Approximate OWL-reasoning with Screech. In: Calvanese, D., Lausen, G. (eds.) RR 2008. LNCS, vol. 5341, pp. 165–180. Springer, Heidelberg (2008)
25. W3C OWL Working Group. OWL 2 Web Ontology Language: Document Overview. W3C Working Draft, September 22 (2009),
http://www.w3.org/TR/2009/PR-owl2-overview-20090922/

A   Appendix: Suggestions for OWL 3

Abstract. With OWL 2 about to be completed, it is the right time to start discussions on possible future modifications of OWL. We present here a number of suggestions in order to discuss them with the OWL user community. They encompass expressive extensions on polynomial OWL 2 profiles, a suggestion for an OWL Rules language, and expressive extensions for OWL DL.

A.1   Introduction

The OWL community has grown with breathtaking speed in the last couple of years. The improvements coming from the transition from OWL 1 [17] to OWL 2 [25] are an important contribution to keeping the language alive and in synch with the users. While the standardization process for OWL 2 is currently coming to a successful conclusion, it is important that the development process does not stop, and that discussions on how to improve the language continue.

In this appendix, we present a number of suggestions for improvements to OWL DL,[9] which are based on some recent work. We consider it important that such further development is done in alignment with the design principles of OWL, and in particular with the description logic perspective which has governed its creation. Indeed, this heritage has been respected in the development of OWL 2, and is bringing it to a fruitful conclusion. There is no apparent reason for straying from this path.

In particular, the following general rationales should be adhered to, as has happened for OWL 1 and OWL 2.

- Decidability of OWL DL should be retained.
- OWL DL semantics should be based on a first-order predicate logic semantics (and as such should, in particular, be essentially open-world and monotonic).
- Analysis of computational complexities shall govern the selection of language features in OWL DL.

Obviously, there are other important issues, like basic compatibility with RDF, having an XML-based syntax, backward-compatibility, etc., but we take these for granted and do not consider them to be major obstacles as long as future extensions of OWL are developed along the inherited lines of thinking.

The suggestions which we present below indeed adhere to the design rationales just laid out. They concern different aspects of the language, and are basically independent of each other, i.e. they can be discussed separately. At the same time, however, they are also closely related and compatible, so that it is reasonable to discuss them together.

---

[9] OWL DL has always played a special role in defining OWL – it is the basis from which OWL Full and other variants, like OWL Lite or the OWL 2 profiles, are developed. So in this appendix we focus on OWL DL.

In Section A.2, we suggest a rule-based syntax for OWL. The syntax is actually of a hybrid nature, and allows e.g. class descriptions inside the rules. Nevertheless, it captures OWL with a syntax which is essentially a rule-syntax.

In Section A.3, we suggest the addition of Boolean role expressions to the arsenal of language constructs available in OWL. We also explain which cautionary measures need to be taken when this is done, in order to not lose decidability and complexity properties.

In Section A.4, we suggest considerably extending OWL by including the DL-safe variable fragment of SWRL [9] together with the DL-safe fragment [21] of SWRL.

In Section A.5, we propose a tractable profile, called ELP, which encompasses OWL 2 EL, OWL 2 RL, most of OWL 2 QL, and some expressive means which are not contained in OWL 2. It is currently the most expressive polynomial language which extends OWL 2 EL and OWL 2 RL, and is still relatively easy to implement.

In Section A.6, we conclude.

Obviously, we do not have the space to define all these extensions in detail, or to discuss all aspects of them exhaustively. We thus strive to convey the main ideas and intuitions, and refer to the indicated literature for details. In the definitions and discussions, we will sometimes drop details, or remain a bit vague (and thus compromise completeness of our exhibition), in order to be better able to focus on the main arguments. We believe that this serves the discussion better than being entirely rigorous on the formal aspects.

A.2   An OWL Rules Language

The alignment of rule languages with OWL (and vice versa) has been a much (and sometimes heatedly) discussed topic. The OWL paradigm is quite different in underlying intuition, modelling style, and expressivity than standard rule language paradigms. Recent efforts involving OWL and rules attempt to merge the paradigms in order to get the best of both worlds.

The advance from OWL 1 to OWL 2 indeed brings the two paradigms closer together. In particular, a considerable variety of rules, understood as Datalog rules with unary and binary predicates under a first-order predicate logic semantics, can be translated with some effort directly into OWL 2 DL. This observation paves the way for a rule-based syntax for OWL, which we will briefly present below. The suggestions in this section are based on [11].

Given any description logic D, a D-rule is a rule of the form

$$A_1 \wedge \cdots \wedge A_n \rightarrow A,$$

where A and A_i are expressions of the form $C(x)$ or $R(x, y)$, where C are (possibly non-atomic) concept expressions over D, R are role names (or role expressions if allowed in D), and x, y are either variables or individual names (y may also be a datatype value if this is allowed in D), and the following conditions are satisfied.

- The pattern of variables in the rule body forms a tree. This is to be understood in the sense that whenever there is an expression $R(x, y)$ with a role

$$\text{Man}(x) \land \text{hasBrother}(x,y) \land \text{hasChild}(y,z) \to \text{Uncle}(x)$$
$$\text{ThaiCurry}(x) \to \exists \text{contains.FishProduct}(x)$$
$$\text{kills}(x,x) \to \text{PersonCommittingSuicide}(x)$$
$$\text{PersonCommittingSuicide}(x) \to \text{kills}(x,x)$$
$$\text{NutAllergic}(x) \land \text{NutProduct}(y) \to \text{dislikes}(x,y)$$
$$\text{dislikes}(x,z) \land \text{Dish}(y) \land \text{contains}^-(z,y) \to \text{dislikes}(x,y)$$
$$\text{worksAt}(x,y) \land \text{University}(y) \land$$
$$\land \text{supervises}(x,z) \land \text{PhDStudent}(z) \to \text{professorOf}(x,z)$$
$$\text{Mouse}(x) \land \exists \text{hasNose.TrunkLike}(y) \to \text{smallerThan}(x,y)$$

Fig. 1. A $\mathcal{SROIQ}$-Rules knowledge base

R and two variables x,y in the rule body, then there is a directed edge from x to y – hence each body gives rise to a directed graph, and the condition states that this graph must be a tree. Note that individuals are not taken into account when constructing the graph.[10] Note also that the rule body must form a single tree.

– The first argument of A is the root of the just mentioned tree.

Semantically, $\mathcal{SROIQ}$-rules come with the straightforward meaning under a first-order predicate logic reading, i.e., the implication arrow is read as first-order implication, and the free variables are considered to be universally quantified.

A D-Rules knowledge base consists of a (finite) set of D-rules,[11] which satisfies additional constraints, which depend on D. These constraints guarantee that certain properties of D, e.g., decidability, are preserved.

For OWL 2 DL, these additional constraints specify regularity conditions and restrictions on the use of non-simple roles, similarly to $\mathcal{SROIQ}(D)$ – we omit the details. Examples for $\mathcal{SROIQ}$-rules are given in Figure 1.

The beauty of $\mathcal{SROIQ}$-rules lies in the fact that any $\mathcal{SROIQ}$-Rules knowledge base can be transformed into a $\mathcal{SROIQ}$ knowlege base – and that the transformation algorithm is polynomial. This means that $\mathcal{SROIQ}$-rules are nothing more or less than a sophisticated kind of syntactic sugar for $\mathcal{SROIQ}$. It is easy to see that, in fact, any $\mathcal{SROIQ}$-axiom can also be written as a $\mathcal{SROIQ}$-rule, so that modelling in $\mathcal{SROIQ}$ can be done entirely within the $\mathcal{SROIQ}$-Rules paradigm.

In order to be a useful language, it is certainly important to develop convenient web-enabled syntaxes. Such a syntax could be based on the Rule Interchange Format (RIF) [1], for example, which is currently in the final stages of becoming a W3C Recommendation. A $\mathcal{SROIQ}$-Rules syntax could also be defined as a straightforward extension of the OWL 2 Functional Style Syntax [20].

---

[10] The exact definition is a bit more complicated; see [11].

[11] Notice the difference in spelling: uppercase vs. lowercase.

$$\exists(\text{testifiesAgainst} \sqcap \text{relativeOf}).\top \sqsubseteq \neg\text{UnderOath}$$
$$\text{hasParent} \sqsubseteq \text{hasFather} \sqcup \text{hasMother}$$
$$\text{hasDaughter} \sqsubseteq \text{hasChild} \sqcap \neg\text{hasSon}$$

Fig. 2. Examples for Boolean role constructors

Proposal: OWL 3 should have a rule-based syntax based on Description Logic Rules.

A.3   Boolean Role Constructors

Boolean role constructors, i.e., conjunction, disjunction, and negation for roles, can be added to description logics around OWL under certain restrictions, without compromising language complexity. Since they provide additional modelling features which are clearly useful in the right circumstances (see Figure 2), there is no strong reason why they shouldn't be added to OWL. The following summarizes results from [22].

All Boolean role constructors can be added to $\mathcal{SROIQ}$ without compromising its computational complexity, as long as the constructors involve only simple roles – the resulting description logic is denoted by $\mathcal{SROIQB}_S$.

Likewise, OWL 2 EL can be extended with role conjunction without losing polynomial complexity of the language. Regularity requirements coming from $\mathcal{SROIQ}$ can be dropped (they are also not needed for polynomiality of the description logic $\mathcal{EL}^{++}$, which is well-known). Likewise, the extension of OWL 2 RL with role conjunctions is still polynomial.

While the complexity results just given are favorable, it has to be noted that suitable algorithms for reasoning with $\mathcal{SROIQB}_S$ still need to be developed. Algorithms for the respective extensions of OWL 2 EL and OWL 2 RL, however, can easily be obtained by adjusting known algorithms for these languages – see also Section A.5.

Proposal: OWL 3 should allow the use of Boolean role constructors wherever appropriate.

A.4   DL-Safe Variable SWRL

SWRL [9] is a very natural extension for description logics with first-order predicate logic rules. Despite being a W3C Member Submission rather than a W3C Recommendation, it has achieved an extremely high visibility. However, in its original form, SWRL is undecidable, i.e., it does not closely follow the design guidelines we have listed in the introduction.

A remedy for the decidability issue is the restriction of SWRL rules to so-called *DL-safe rules* [21]. Syntactically, DL-safe rules are rules of the form

$$A_1 \wedge \cdots \wedge A_n \rightarrow A$$

as in Section A.2, but without the requirements on tree-shapedness. Semantically, however, they are read as first-order predicate logic rules, but with the restriction that variables in the rules may bind only to individuals which are present in the knowledge base.[12] In essence, this limits the usability of DL-safe SWRL to applications which do not involve TBox reasoning.

It is now possible to generalize DL-safe SWRL without compromising decidability. The underlying idea has been spelled out in a more limited setting in [12] (see also Section A.5), but it obviously carries over to $\mathcal{SROIQ}$.

In order to understand the generalization, we need to return to $\mathcal{SROIQ}$-rules as defined in Section A.2. Recall that the tree-shapedness of the rule bodies is essential, but that role expressions involving individuals are ignored when checking for the tree structure.

The idea behind DL-safe variable SWRL is now to identify those variables in rule bodies which violate the required tree structure, and to define the semantics of the rules such that these variables may only bind to individuals present in the knowledge base – these variables are called *DL-safe variables*. The other variables are interpreted as usual under the first-order predicate logic semantics.

An alternative way to describe the same thing is to say that a rule qualifies as DL-safe variable SWRL if replacing all DL-safe variables in the rule by individuals results in an allowed $\mathcal{SROIQ}$-rule.

As an example, consider the rule

$$C(x) \wedge R(x,w) \wedge S(x,y) \wedge D(y) \wedge T(y,w) \to V(x,y),$$

which violates the requirement of tree-shapedness because there are two different paths from x to w. Now, if we replace w by an individual, say o, then the resulting rule

$$C(x) \wedge R(x,o) \wedge S(x,y) \wedge D(y) \wedge T(y,o) \to V(x,y)$$

is a $\mathcal{SROIQ}$-rule.[13] Hence, the rule

$$C(x) \wedge R(x,w_s) \wedge S(x,y) \wedge D(y) \wedge T(y,w_s) \to V(x,y),$$

---

[12] The original definition is different, but equivalent. It required that each variable occurred in an atom in the rule body, which is not an atom of the underlying description logic knowledge base. The usual way to achieve this is by introducing an auxiliary class O which contains all known individuals, and adding $O(x)$ to each rule body, for each variable in the rule. Our definition instead employs a redefinition of the semantics, which appears to be more natural in this case. Essentially, the two formulations are equivalent.

[13] This rule can be expressed in $\mathcal{SROIQ}$ by the knowledge base consisting of the three statements

$$C \sqcap \exists R.\{o\} \sqsubseteq \exists R_1.\mathrm{Self}$$
$$D \sqcap \exists T.\{o\} \sqsubseteq \exists R_2.\mathrm{Self} \qquad \text{and}$$
$$R_1 \circ S \circ R_2 \sqsubseteq V.$$

See [11].

where w_s is a DL-safe variable, is a DL-safe variabe SWRL rule. Note that the other variables can still bind to elements whose existence is guaranteed by the knowledge base but which are not present in the knowledge base as individuals, which would not be possible if the rule were interpreted as DL-safe.

In principle, naive implementations of this language could work with multiple instantiations of rules containing DL-safe variables, but no implementations yet exist. In principle, they should not be much more difficult to deal with than DL-safe SWRL rules.

Proposal: OWL 3 DL should incorporate DL-safe SWRL and DL-safe variable SWRL.

A.5   Pushing the Tractable Profiles

The OWL 2 Profiles document [18] describes three designated profiles of OWL 2, known as OWL 2 EL, OWL 2 RL, and OWL 2 QL. These three languages have been designed with different design principles in mind. They correspond to different description logics, have different expressive features, and can be implemented using different methods.

The three profiles have in common that they are all of polynomial complexity, i.e., they are rather inexpressive languages, despite the fact that they have already found applications. While having three polynomial profiles is fine due to their fundamental differences, the question about maximal expressivity while staying in polynomial time naturally comes into view.

The ELP language [12] is a language with polynomial complexity which properly contains both OWL EL and OWL RL. It also contains most of OWL QL.[14] Furthermore, it still features rather simple algorithms for reasoning implementations.

More precisely, ELP has the following language features.

- It contains OWL 2 EL Rules, i.e. $\mathcal{EL}^{++}$-rules as defined in Section A.2.[15] Note that $\mathcal{EL}^{++}$-rules cannot be converted to $\mathcal{EL}^{++}$ (i.e. OWL 2 EL) using the algorithm which converts $\mathcal{SROIQ}$-rules to $\mathcal{SROIQ}$.
- It allows role conjunctions for simple roles.
- It allows the use of DL-safe variable SWRL rules, in the sense that replacement of the safe variables by individuals in a rule must result in a valid $\mathcal{EL}^{++}$-rule.
- General DL-safe Datalog[16] rules are allowed.

The last point – allowing general DL-safe Datalog rules – is a bit tricky. As stated, it destroys polynomial complexity. However, if there is a global bound on the number of variables allowed in Datalog rules, then polynomiality is retained. Obviously, one would not want to enforce such a global bound; nevertheless

---

[14] Role inverses cannot be expressed in ELP.

[15] $\mathcal{EL}^{++}$-rules are D-rules with $D = \mathcal{EL}^{++}$.

[16] One could also simply allow DL-safe SWRL rules.

$$\text{NutAllergic}(x) \land \text{NutProduct}(y) \rightarrow \text{dislikes}(x,y)$$
$$\text{Vegetarian}(x) \land \text{FishProduct}(y) \rightarrow \text{dislikes}(x,y)$$
$$\text{orderedDish}(x,y) \land \text{dislikes}(x,y) \rightarrow \text{Unhappy}(x)$$
$$\text{dislikes}(x,v_s) \land \text{Dish}(y) \land \text{contains}(y,v_s) \rightarrow \text{dislikes}(x,y)$$
$$\text{orderedDish}(x,y) \rightarrow \text{Dish}(y)$$
$$\text{ThaiCurry}(x) \rightarrow \text{contains}(x,\text{peanutOil})$$
$$\text{ThaiCurry}(x) \rightarrow \exists\text{contains.FishProduct}(x)$$
$$\rightarrow \text{NutProduct}(\text{peanutOil})$$
$$\rightarrow \text{NutAllergic}(\text{sebastian})$$
$$\rightarrow \exists\text{orderedDish.ThaiCurry}(\text{sebastian})$$
$$\rightarrow \text{Vegetarian}(\text{markus})$$
$$\rightarrow \exists\text{orderedDish.ThaiCurry}(\text{markus})$$

Fig. 3. A simple example ELP rule base about food preferences. The variable v_s is assumed to be DL-safe.

the result indicates that a careful and limited use of DL-safe Datalog rules in conjunction with a polynomial description logic should not in general have a major impact on reasoning efficiency.

Since ELP is fundamentally based on $\mathcal{EL}^{++}$-rules, it features rules-style modelling in the sense in which $\mathcal{SROIQ}$-rules provide a rules modelling paradigm for $\mathcal{SROIQ}$. An example knowledge base can be found in Figure 3.

As for implementability, reasoning in ELP can be done by means of a polynomial-time reduction to Datalog, using standard Datalog reasoners. Note that TBox-reasoning can be emulated even if the Datalog reasoner has no native support for this type of reasoning. The corresponding algorithm is given in [12]. An implementation is currently under way.

Proposal: OWL 3 should feature a designated polynomial profile which is as large as possible, based on ELP.

A.6   Conclusions

Following the basic design principles for OWL, we made four suggestions for OWL 3.

- OWL 3 should have a rule-based syntax based on Description Logic Rules.
- OWL 3 should allow the use of Boolean role constructors.
- OWL 3 should incorporate DL-safe SWRL and DL-safe variable SWRL.
- OWL 3 should feature a designated polynomial profile which is as large as possible, based on ELP.

We are aware that these are only first suggestions, and that a few open points remain to be addressed in research. We hope that our suggestions stimulate discussion which will in the end lead to a favorable balance between application needs and language development from first principles.

A Functional Ontology of Observation and Measurement

Werner Kuhn

Institute for Geoinformatics, University of Muenster
Weselerstr. 253, 48151 Münster, Germany
kuhn@uni-muenster.de

Abstract. An ontology of observation and measurement is proposed, which models the relevant information processes independently of sensor technology. It is kept at a sufficiently general level to be widely applicable as well as compatible with a broad range of existing and evolving sensor and measurement standards. Its primary purpose is to serve as an extensible backbone for standards in the emerging semantic sensor web. It also provides a foundation for semantic reference systems by grounding the semantics of observations, as generators of data. In its current state, it does not yet deal with resolution and uncertainty, nor does it specify the notion of a semantic datum formally, but it establishes the ontological basis for these as well as other extensions.

Keywords: observation, measurement, ontology, semantics, sensors.

1 Introduction

Given that observation is the root of information, it is surprising how little we understand its ontology. Measurement theory, the body of literature on the mathematics of measurements, is only representational, treating questions of how to represent observed phenomena by symbols and how to manipulate these. Ontological questions like „what can be observed" or „how do observations relate to reality?" are not answered by it. Consequently, the semantics of information in general, and of observations in particular, rests on shaky ground. With sensor observations becoming ubiquitous and major societal decisions (concerning, for example, climate, security, or health) being taken based on them, an improved understanding of observations as information has become imperative. Answering some of the deepest and most pressing questions in geographic information science, such as how to model and monitor change, also requires progress in this direction. Furthermore, issues of scale, quality, trust, and reputation, are all intimately linked to observation processes.

In response to these needs, this paper proposes a first cut at an ontology of observation. The ontology specifies observation as a process, not only as a result, and treats it as an information item with semantics that are independent of observation technology. The goal of this work is to understand the information processes involved in observations, not the details of physical, psychological, or technological processes.

K. Janowicz, M. Raubal, and S. Levashkin (Eds.): GeoS 2009, LNCS 5892, pp. 26–43, 2009.

The ontology is kept at a sufficiently general level to be widely applicable as well as compatible with a broad range of existing methods and standards. Its primary purpose is to serve as a backbone for a seamless semantic sensor web. It also provides a foundation for semantic reference systems, systematizing the semantics of observations, which are the basic elements of spatial information. More generally, it serves as an ontological account of data production.

Observations link information to reality and provide the building blocks of conceptualizations. As such they ground communication and relate data to the world and its observers. The paper shows that they hold together the four top-level branches in the foundational DOLCE ontology [1]: they are afforded by changes in the environment (stimuli), which involve endurants and perdurants, and their results consist of abstract symbols, which stand for qualities inhering in these endurants and perdurants. Studying the ontology of observation is, thus, also likely to clarify fundamental categories in ontology as well as their mutual interaction.

The proposed ontology is presented in the form of a simulation of observation processes, providing a testable model as an additional benefit. To achieve this, it is written in the typed functional language Haskell [2], providing the expressiveness required to capture and distinguish concepts like endurants, perdurants, qualities, qualia, stimuli, signals, and values. The ontology is built and tested by specifying algebraic theories and models for these concepts. Its development combines a bottom-up strategy, working from actual use cases, with a top-down structure, taken from DOLCE. It does not force the expressive limitations of semantic web languages and reasoners on the modeling and representation of observations, though it allows for subsequent translation into these representations.

The technological focus of today's sensor standards, putting encoding before modeling, motivates an effort to „lift sensors from their platforms", so to speak. The proposed ontology generalizes from technical and human sensors [3] to the role of an observer. It models observation as an action afforded to observers by their environment. People, devices, sensor systems and sensor networks can then all realize this affordance, leading to a vast generalization of sensing behavior, which simplifies software architectures [4, 5].

The paper reviews previous work relevant to the ontology of observation (section 2), states the ontological commitments taken and their implications (section 3), presents the core concepts of observation (section 4) followed by a series of examples (section 5), and walks the reader through the observation ontology and its formalization (section 6), before ending with conclusions and an outlook (section 7).

2 Previous Work

Observation and measurement processes have received attention from several perspectives, including physics, mathematics and statistics, ontology, and information sciences. They also have been the subjects of recent and ongoing standardization efforts. Yet, the ontology of observation processes remains surprisingly underdeveloped. The focus is typically on endurants (in particular, physical objects) rather than on perdurants and qualities and we lack an understanding of observation in general, as opposed to specific kinds of observations in various application areas.

This section reviews some results from science and engineering that are relevant to a more general observation ontology. The broader topic of the ontology of qualities is beyond the scope of this paper and will only be touched upon where it is necessary to understand the ontology of observation. The narrower problem of the ontology of measurement units is factored out, as it represents the focus of parallel research and standardization activities. The question whether there are observations that are ontologically more basic than others is also not addressed here.

2.1 Physics

Physics (and metaphysics) has asked questions about the nature of observation all the way back to (at least) Aristotle. Relevant core ontological distinctions have resulted from these analyses. Modern physics has revealed challenges (like the interaction of the observer with the observed and the limitations to precise knowledge) that affect all theorizing and some of the practice of observing. In his insightful book *Physics as Metaphor*, Roger S. Jones [6] pointed out that consciousness plays a much larger role in constructing physical reality than commonly accepted and that the separability of mind and matter is untenable even from the point of view of measurement. He argued that "the celebrated ability to quantify the world is no guarantee of objectivity and that measurement itself is a value judgment created by the human mind." Basic measurements like that of length are not well defined, according to Jones, but have a built-in unspecifiable uncertainty, in the case of length due to the circularity of demarcating what is being measured using length itself (p. 22). While one can take different stances on such philosophical underpinnings of observation and measurement, it is clear that only a careful ontological specification will make these stances explicit and testable.

2.2 Mathematics and Statistics

Mathematics and statistics have addressed measurement from a representational point of view: what properties do symbols need to have to represent observations and how can they be manipulated to reveal statistical properties? [7]. This led to the well-known measurement scales, i.e. classifications of measurement variables according to their algebraic properties. Statistical views of measurement are largely orthogonal to ontological perspectives. Measurement theory does ask ontological questions about measurement scales (regarding, for example, the presence of order relations or of an absolute zero), but the answers to these questions are derived from analytical needs rather than ontological analyses of observed phenomena. Thus, it remains up to a data analyst, for example, to assign the scale levels requested by statistics or visualization programs.

Stevens' measurement scales, with their underdeveloped link to ontology, sometimes get replaced by representational distinctions, such as continuous, discrete, categorical, and narrative attributes [8]. While this is useful for assigning probability distributions, it loosens the connection to what is being measured. As a consequence, the assumptions made in statistics about observed variables are sometimes at odds with ontological analysis. This is not only the case for the assigned measurement scales, but, more fundamentally, for the claims that a measurement variable (like the temperature in a room) has a „true" value. Since the „truth" of this value is defined by the observation procedure (involving sampling processes, spatial and temporal resolutions, interference with the measured phenomenon etc.), any procedure has its own „truth". Presupposing a true measurement scale or a true value of a measurement variable begs the ontological questions about what is being measured and how (see also [9]). When observations from different sources, holding different „truths", are combined, such discrepancies show up at best (and then need to be accounted for post hoc, often without the necessary information) or they get absorbed in biased statistical measures (by inadvertently mixing multiple sampling processes).

2.3 Ontology

A thorough ontological analysis of measurements has been proposed in [10]. It provides the first explicit specification of measurement quantities and engineering models in the form of a formalized ontology. Its focus is on the core distinctions of scalar, vector, and tensor quantities, physical dimensions, units of measure, functions of quantities, and dimensionless quantities. As such, the ontology isolates the measured quantities (expressed by symbols and units) from the measurement process and from the bearers of the measured qualities. This decoupling of concerns is entirely valid and supports a combination with ontologies of measured entities and measurement processes.

Again from an engineering perspective, [11] has proposed a taxonomic approach to sensor science. Their scope includes the physics and technology of measurement, and their taxonomy can be seen as a weak form of an ontology, coming close to the information centric view of sensing and measurement advanced here. They also make the case for a unified treatment of technical and human sensors, and point out that sensor science has (at least at that time) not placed enough emphasis upon the sensing function.

The measurement ontology proposed in [12] is meant to extend to aspects of measurement left out in previous attempts, such as sampling and quality evaluation. It emphasizes the use of measurements more than their semantics and takes a simplified view of measurement as a function from an object to a numerical value. The deeper question about how a measurement result relates to an object (if any) is left open. The ontology is based on an ontology engineering method, but not on a foundational ontology. It introduces some complex notions (like "traceable resource units"), which are difficult to understand and to relate to other ontologies.

A recent proposal for ontological extensions to sensor standards [13] rests on representational distinctions from [14] without developing or refining them further. An ontological analysis of observations and measurements that goes beyond an RDF or UML level representation of XML data types is still lacking and proposed in this paper.

2.4 Geographic Information Science

Geographic Information Science has paid considerable attention to fundamental questions about the nature of information and observation in space and time. Chrisman [15] was the first to call for attribute reference systems, to complement spatial and temporal reference systems in supporting the interpretation of data. This was essentially a call for ontologies of observation and measurement. But Chrisman's notion of reference system remained anchored in measurement scales and their extensions, without providing a theoretical basis for assigning the scales in the first place or an ontological account of measurands.

A generalized notion of semantic reference systems, encompassing space, time, and theme, and establishing a link to ontologies was proposed in [16, 17]. This research program raises the question of how to define a semantic datum and how to ontologize observations, as the basic components of geodata. Probst [18] provided the first definition of a semantic datum and advanced the ontology of qualities, in particular of spatial qualities, which he related to the dimensions of their carriers [19]. Building on the notion of qualities from [20], he formalized DOLCE's quality spaces (which are in turn based on Gärdenfors' conceptual spaces [21]) and introduced reference spaces as quality spaces that have been partitioned by symbols denoting measurement values. Schade has recently extended this work to enable semantic translation between attribute values [22].

The questions how measured qualities relate to their carriers remained open in this line of work, apart from the dimensional restrictions identified in [19]. In a recent paper [23], we have presented a first account of how observations relate to endurants and perdurants, leading to a revised and simpler definition of a semantic datum based on Gibson's ecological psychology [24]. The present paper complements this account with an ontological analysis of observations and measurements from the point of view of their acquisition.

From a data model perspective, efforts have been undertaken to anchor geodata in fundamental models of observation, typically based on the physics notion of fields (for the latest example, see [25]) and then to provide ontological accounts of subsequent abstraction levels [26, 27]. These efforts typically presuppose absolute space and time and introduce the convenient abstraction of point observations, which can be seen as a surrogate ontology of observation. Given their different scopes (data models on the one hand, more general geographic information ontologies on the other), they are compatible with the approach presented here. They differ from it by tying observation to location (e.g., a measure of temperature at some point in a space-time continuum) instead of to endurants and perdurants (e.g., a measure of temperature of the amount of air surrounding a thermometer).

From a data processing perspective, measurement-based systems have been proposed for geospatial information [28, 29], with the purpose of establishing a foundation for spatial analysis through the collection of maximally original data. Positions and their uncertainties, for example, can more reliably and even incrementally be determined if the original terrestrial or satellite measurements are maintained in a system. A generalization of measurement-based approaches toward models that trace their concepts back to measurements requires an ontology of measurement. This will allow for data integration and for computational techniques to be applied to collections of measurements (such as incremental least-squares adjustment [30, 31]). One could even argue that the limited take-up of the idea of measurement-based systems until now has to do with its lacking ontological foundation, preventing a link between measurements and other data about objects and processes.

Recently, Goodchild has proposed the notion of Volunteered Geographic Information (VGI), and the idea of Citizens as Sensors going hand in hand with it [3]. Such a conceptualization of sensors as roles played by machines as well as humans lies at the heart of the approach presented here, which has the practical advantage of turning the observation ontology into a solid foundation for volunteered information, and more generally for the social web.

2.5 Standardization

The ontological ambiguities inherent in current sensor and observation standards prevent an integration of observation data from multiple sources and their appropriate interpretation in models. The envisioned Semantic Sensor Web [32, 33] needs a stronger ontological foundation to become reality, as evidenced by today's lack of access to sensor data on the web.

Broadly speaking, three aspects of observation and measurement have been the subjects of standardization so far:

- measurement *units* (leading to the SI system of measurement units as well as some ontologies of measurement units);
- measurement *uncertainty* (leading to standard ways of applying probability theory and statistics to measurements as well as early standardization efforts on exchange of data on uncertainty);
- measurement *technology* (leading to various standards of instrumentation and communication).

Where such standardization efforts are not based on an ontology, problems can arise in using and combining them. For example, the Observations and Measurements standard of the Open Geospatial Consortium (OGC, [14]) provides a model for describing and XML schemas for encoding observations and measurements, with the goal of supporting interoperability. However, it states that all observed properties are properties of "features of interest", which it defines as "the real-world object regarding which the observation is made" [14]." Apart from the notorious confusion between real world and information objects, this restricts observations unnecessarily and inconveniently to properties of objects. Where no such object can be identified (e.g., for weather data), even a sensor can become the „feature of interest". Probst [34] has shown some contradictions and interoperability impediments arising from such ontologically unfounded standards and showed how to remedy the situation, for example by treating the feature of interest as a role.

The work presented here extends this idea further, to treat other observation concepts as roles as well, in particular sensors and stimuli. Sensors are modeled from an information-centric point of view, rather than from the technological and information encoding perspective of geospatial standards. A recent effort toward a sensor ontology [35] takes a similar direction, though it remains closely tied to technology.

3 Ontological Commitments

The ontology of observation is hindered by, among other factors, the naive idea of a measurement instrument being an objective reporter of the mind-independent state of the world. This commonly held view neglects the fact that instruments are built and calibrated by human beings. Neither the choice of the observed entity, nor the quality assigned to it, nor its link to a stimulus, nor the value assigned to the quality are mind-independent. All of them involve human conceptualizations, though these are more amenable to grounding and agreement than anything else. Therefore, an ontology of observation requires ontological commitments of one sort or another and those taken for this work are spelt out in this section.

As ontological foundation, DOLCE's distinction of four top level categories of particulars is adopted here [1]: endurants, perdurants, abstracts, and qualities (which can be physical, temporal, or abstract). Endurants, for example lakes, participate in perdurants, for example rainfalls. The categorization of an entity as endurant or perdurant is often a matter of the desired temporal resolution. On closer analysis, many phenomena involve both categories. For example, a water body can be conceptualized as an endurant, neglecting the flow of water, or as a mereological sum of endurants (amounts of water, terrain features) participating in a water flow perdurant.

Qualities inhere in particulars and map them to abstracts (regions in quality spaces). Physical qualities inhere in physical endurants, temporal qualities in perdurants. For example, a temperature quality inheres in an amount of matter and a duration quality inheres in an event. Quality universals shall be admitted, so that a quality can be abstracted from multiple instances to a quality type (e.g., air temperature) and even further to a generalized quality type (e.g., temperature).

As in [18], an observation process is seen here as invoking first a quale in the observer's mind, or an analog signal in a technical sensor. Our notion of qualia is slightly different from the one in [20], but in line with the one in philosophy of mind (for a good overview, see http://en.wikipedia.org/wiki/Qualia): it denotes a quality experienced by an observer and is not abstracted from the carrier of the quality. The red of the rose experienced by an observer belongs to that particular rose as well as to the observer; it is not abstracted from either. Thus, observation involves firstly the production of a quale (analog signal) and secondly its symbolization, i.e., a sequence of impression and expression.

Measuring is distinguished here from observing by requiring measurements to have numeric results. Stevens considered measurement as „the assignment of numerals to objects and events according to rule" (p. 677 of [7]), but included names in his measurement scales. This required, at least in theory, to turn names into numbers by some rules. Here, the term measurement is restricted to quantification, and the term observation is used for sensing processes with results symbolized in any form, not just numerically. The defining property of both, observation and measurement, is that they map qualia to well defined symbol algebras, whether these are numeric or not.

The result of an observation (as well as of a measurement) process is an information object, which is commonly referred to as observation as well. It is a non-physical endurant, expressed by abstract symbols. Apart from the value of the observed quality, it can contain temporal and location as well as uncertainty information. Our ontology uses the terms observe for the process and Observation for its result.

4 Core Observation Concepts

This section defines the core concepts of observation, using the ontological foundation introduced in the previous section: the notions of observable, stimulus, observer, observation value, and observation process. The guideline for these choices has been to remain as compatible as possible with existing standards, with the literature (such as [36]) and with ordinary language. For example, the common distinction between observation and measurement (the latter having a numeric result) has been retained, and the term sensor denotes technical devices, while the term observer is used for the generalization over humans and devices. On the other hand, ambiguous terms in current sensor standards have been made more precise. For example, sensors have no knowledge of objects and therefore cannot observe object properties. The use of the term observation for both, the process of observing and the result is common and should not cause confusion.

An **observable** is a physical or temporal quality to be observed. For example, the temperature of an amount of air or the duration of an earthquake. Ontologically, an observable combines the quality with the entity it inheres in. If the quality to be measured is temperature, the physical endurant it inheres in can be any amount of matter (air, water, etc.) holding heat energy. The choice of a quality bearing endurant (say, of an air mass surrounding a thermometer) determines the spatial resolution of an observation. If the quality is duration, the perdurant it inheres in can be any event, such as an earthquake, a chess game, or the reign of a dynasty.

Since observables per se cannot be detected (how would information about them enter the observer?), the idea of a **stimulus** is needed, explaining how a signal (and eventually, information) is generated. A stimulus is defined in physiology as a „detectable change in the internal or external environment" of an observer (http://en.wikipedia.org/wiki/Stimulus_(physiology)) or as a "physical or chemical change in the environment that leads to a response controlled by the nervous system" (http://www.emc.maricopa.edu/faculty/farabee/BIOBK/BioBookglossS.html). Simple examples of stimuli are the heat energy flowing between an amount of air and a thermometer or the seismic waves of an earthquake. Stimuli need to have a well-defined physical or chemical relationship to observables. A detectable change is a perdurant and can be a process (periodic or continuous) or an event (intermittent), playing the role of a stimulus when an observer detects it. An observer can also produce the necessary stimulus itself (e.g., a sonar producing a sound wave to measure distance). The stimulus can itself be an observation process (changing the observed value). This recursion allows for observations to combine individual observation results into symbols representing aggregate qualities.

Detecting a stimulus requires that an endurant in the internal or external environment of an observer participates in the stimulus. For example, heat flow can be detected by an amount of gas expanding in a thermometer. When the measurand inheres in a perdurant rather than an endurant, there still needs to be a participating endurant; for example, an inertial mass in a seismometer, to be moved by seismic waves. Thus, to be detectable, a stimulus needs to provide one of the following:

- a changing physical quality of a participating endurant (e.g., the volume of gas);
- a temporal quality of the stimulus or of a perdurant coupled with it (e.g., the duration of the motion of an inertial mass);
- an abstract quality of an observation result (e.g., a temperature value).

An **observer** provides a symbol for a quality, in two steps. First, it detects

- a quality of a stimulus or
- one or more proxy qualities (also known as signal variables) of endurants or perdurants internal to the observer.

Proxy qualities have to co-vary with the observable in a well-defined way, either through a participation of their carrier endurants in the stimuli or through a process coupling between their carrier perdurants and the stimuli. Second, the observer expresses the analog signal(s) obtained this way through a symbol for the **value** of the observation. This value is either a Boolean, a count, a measure with a unit, or a category (as in [37]) and can get „stamped" with the position and time of the observation. The observer role can be played by devices (technical sensors) or humans or animals, either individually or in groups.

The **observation** process can now be conceptualized as consisting of the following steps (the first two required only once, to determine the observed phenomenon):

1. choose an observable;
2. find one or more stimuli that are causally linked to the observable;
3. detect the stimuli, producing analog signals ("impression");
4. convert the signals to observation values ("expression").

This sequence contains the ontologically significant elements influencing the semantics of observation values. It is consistent with the definitions of observations in standards for sensors and for geographic information, such as OGC's Observations and Measurements standard ("An Observation is an action with a result which has a value describing some phenomenon." [14]) or OGC's Reference Model ("An observation is an act associated with a discrete time instant or period through which a number, term or other symbol is assigned to a phenomenon." [38]).

5 Observation Examples

This section lists examples of technical and human observers taking a wide variety of observations. It illustrates the notions introduced in section 4 and provides the test cases with which the ontology has been developed and tested. Each example identifies an observer, observable, stimulus, and value.

A **thermometer** measures the temperature of an amount of air using heat flow as a stimulus. The stimulus causes an expansion of an amount of gas, the amount of which (relative to its container) is the signal that gets converted to a number of degrees on the Celsius scale.

A **sonar** measures water depth on a lake using sound waves it generates as stimulus and converting the time until they return from the ground (signal) into a measure of distance.

A **CCD camera** observes its visible environment using sunlight reflected from the surfaces in the environment as stimulus. It integrates the received radiation intensity at each of its pixels over some time interval (signal), and returns an image as observation.

A **weather station** reports the observable "type of weather" by combining temperature, pressure, and humidity measurements (each of them a stimulus producing a signal) and aggregating their values.

A **sailor** observes wind speed by watching the frequency and size of ripples on the sea as stimulus and reporting a Beaufort number expressing his impression (quale) of the wind force.

A **nomad** in the desert reports the presence of water in a well by observing sunlight reflected from patches of water (stimulus), getting the impression (quale) that there is some water, and calling a number on a cell phone signifying „water available" [39].

An **epidemiologist** collects data on dengue fever risk by separating and counting mosquito eggs in a bucket (their presence being the stimulus, their number the observable, their individuation the qualia).

A **doctor** observes a patient's mood by talking to the patient and describing her impressions (qualia) obtained from the patient's behavior, which serves as stimulus.

6   A Walk through the Observation Ontology

The observation ontology presented in this section takes the form of an algebraic specification. It is an ontology, because it specifies observation concepts axiomatically; at the same time, it is a simulation, because it has an executable model that can be used for testing. The section walks the reader through the formalization, explaining its form and contents together. The full ontology is available from http://musil.uni-muenster.de/publications/ontologies/.

The software engineering technique of algebraic specification [40] uses many-sorted algebras to specify conceptualizations. These consist of sets of values from multiple sorts with associated operations. Logical axioms, in the form of equations over terms formed from these operations, constrain the interpretation of the symbols. For example, an axiom in an algebraically specified ontology might state that the time stamp of an observation should be interpreted as the mid point of an observation period.

For reasons of expressiveness and ease of testing, the functional language Haskell is used here as an algebraic specification language. It offers a powerful development and testing environment for ontologies, without the restriction to subsets of first order logic and binary relations typical for ontology languages. In particular, it allows for specifying the behavior of observing as a relation over observer, observed entity, and quality types. Introductions to Haskell as a programming language, together with interpreters and compilers, can be found at http://www.haskell.org. An introductory text and language reference is [2], accessible at http://book.realworldhaskell.org/read/. All development and testing of the observation ontology has been done using the interpreter coming with the Glasgow Haskell Compiler, ghci.

6.1   Data Types for Universals

By considering types as theories [41], with operation signatures defining the syntax and equational axioms defining the semantics for a vocabulary, one can write theories of intended meanings, i.e., ontologies, in Haskell. Universals (a.k.a. categories, classes or concepts) are modeled as data types and individuals (a.k.a. instances) as values. Universals are not just flat sets of individuals, but algebraic structures with characteristic operations (a.k.a. methods or relations).

Let us declare a type symbol for each universal (e.g., Person), a function type for each kind of process (e.g., constructing a value for Person from an Id and Position value), and equations on them (e.g., stating that the position of a person remains the one at construction time until it gets changed by a move operation). The Haskell syntax for type declarations uses the keyword **data** (Haskell keywords are boldfaced throughout the paper) followed by a name for the type and a right-hand side introducing a constructor function for values, possibly taking arguments.

A simple example is the declaration of data types for the universals Person and WeatherStation with constructor functions of the same name and two arguments for a name (Id) and a position, which are both data types as well:

```
data Person = Person Id Position
```

```
data WeatherStation = WeatherStation Id Position
```

Type synonyms can be declared using the keyword **type**. For example, the type Id is declared as a synonym of the predefined type String:

```
type Id = String
```

The Position type (not shown here) contains alternative constructors for fixed and mobile positions, with the former encoded as coordinates (with a reference system) or as a toponym.

A core universal in our ontology is that of observation values. Its various constructors are separated by "|" and take more basic types (Bool etc.) as arguments:

```
data Value = Boolean Bool | Count Int |
             Measure Float Unit | Category String
```

An observation consists of an observation value combined with a position and time:

```
data Observation = Observation Value Position ClockTime
```

ClockTime is a predefined Haskell type for system clock time.

An example of a universal for a quality carrying endurant is the type AmountOfAir. It takes two arguments here, one for the amount of heat energy and the other for the amount of moisture it contains (implying that each amount of air has heat and moisture):

```
data AmountOfAir = AmountOfAir Heat Moisture
```

With a definition of the heat and moisture parameters, one can then define individual values for example as follows:

```
muensterAir = AmountOfAir 10.0 70.0
```

Additional data types specifying endurants like other sensor types, quality carriers or measurement units are defined in the full code.

6.2   Type Classes for Behavior and Subsumption

By organizing categories along the subsumption relation, one can transfer behavior from super- to sub-categories. Standard ontology languages define subsumption in terms of instantiation: persons are agents if every person is also an agent. Haskell does not permit this instantiation of an individual to multiple types (because every value has exactly one type), but offers a more powerful form of subsumption, using type classes. These are sets of types sharing some behavior. Type classes are named with upper case letters here, to distinguish them visually from types, and are followed by a parameter for the types belonging to the class (using the same name in lower case):

```
class ENDURANTS endurant
```

Sub-categories are derived from their super-categories using a so-called context (=>):

```
class ENDURANTS physicalEndurant
    => PHYSICAL_ENDURANTS physicalEndurant

class PHYSICAL_ENDURANTS amountOfMatter
    => AMOUNTS_OF_MATTER amountOfMatter

class PHYSICAL_ENDURANTS physicalObject
    => PHYSICAL_OBJECTS physicalObject
```

Behavior can be added at any level of such a class hierarchy. For example, the ability of agents to tell their position distinguishes the class of agents (called agentive physical objects here, APOS) and then gets passed on to derived classes:

```
class PHYSICAL_OBJECTS apo => APOS apo where
    getPosition :: apo -> Position
```

To state that persons and weather stations are agentive physical objects and inherit the behavior of these, the Person and WeatherStation types are declared instances of the APOS class, specifying how each of them realizes the getPosition behavior:

```
instance APOS Person where
    getPosition (Person iD pos) = pos

instance APOS WeatherStation where
    getPosition (WeatherStation iD pos) = pos
```

Note that Haskell's **instance** relation is one between a *type* and a *type class*. Types can instantiate classes (with the same context syntax), without creating dubious cases of multiple is-a relations. An example is the OBSERVERS class introduced below, combining behavior of agents and qualities. Type classes furthermore allow for inheritance, so that penguins can be birds without flying behavior, or some APOS types (e.g., sensors without locating capacity) can be declared unable to tell their position.

6.3   Constructor Functions for Qualities

Qualities are the subject of observation. Each observable quality inheres in an endurant or perdurant, i.e., is a dependent entity. This suggests a specification of quality types as functions and, more specifically, as data constructor functions (constructing values of a certain quality from individual endurants or perdurants). Individual quality values are then the results of applying a quality constructor to an endurant or perdurant. Note that this does not imply that endurants or perdurants be individuated prior to observation processes, only as parts of them.

Quality types can be generalized even further, abstracting the type they inhere into a type parameter. For example, the temperature quality type is specified independently of the kind of physical endurant it describes, using a class context to constrain its parameter:

```
data PHYSICAL_ENDURANTS physicalEndurant
   => Temperature physicalEndurant =
        Temperature physicalEndurant
```

We have seen data constructors before, for example the function `Person`, which takes an Id and a Position value and constructs a value of type `Person`. What is different here is that the constructor can take a value of several types, satisfying the context (i.e., belonging to the type class `PHYSICAL_ENDURANTS`). The declaration establishes a type template, which generates different quality types for different bearer parameters. The temperature of air and that of water, for example, have two signatures with the same constructor function (`Temperature`), but different parameter types. Air temperature can then be specialized through a type synonym:

```
type AirTemperature = Temperature AmountOfAir
```

Following DOLCE, an individual quality was defined in section 3 as a region in a quality space. It is specified here by the term formed by applying a quality constructor to a particular. The following term specifies the air temperature at Münster (using the term `muensterAir` specified above):

```
Temperature muensterAir
```

For a quality of an endurant or perdurant that is internal to a human observer, this term can be seen as representing a quale (e.g. "`Temperature myBody`"). For a technical sensor, it specifies the analog signal generated by the stimulus. In both cases, the term stands for the result of the first step of an observation (i.e., a sense impression), preceding its symbolization in an observation value (i.e., an expression). There is no need and no possibility to evaluate the term further. Its symbolization is specified in the following subsection.

6.4 The Observer Role

Putting it all together, let us now specify the role of an observer, as a class of three kinds of types: the observing agent, the observed quality, and the entity bearing the quality. This so-called multi-parameter type class can be seen as a relation over its types, defining which agents can observe which qualities of which entities. It is characterized by the observe behavior, which uses the `express` operation to symbolize the observed value:

```
class (APOS agent, QUALITIES quality entity)
   => OBSERVERS agent quality entity where
```

```
observe :: quality entity -> agent -> IO Observation
express :: quality entity -> agent -> Value
```

The observe operation feeds a quality of an entity to an observing agent to produce an observation. Since this involves input from the system clock (to time stamp the observation), one can use the Haskell IO Monad to wrap the result of these two operations. Because the observe behavior is the same for all agent and quality types, it can already be implemented in the class specification. Its specification uses the Haskell **do** notation, which allows for sequencing the execution of operations. In a first step, the clock time is read, then the Observation is returned as a triple of the result of the express operation, the position of the agent at the time, and the clock time:

```
observe quale agent = do
   clockTime <- getClockTime
   return (Observation (express quale agent)
                       (getPosition agent) clockTime)
```

Any type of sensing agent that should play the role of an observer can now be instantiated to this type class. All that is left to specify in such instantiations is how the agent symbolizes observations (using the express operation). For example, the ability of weather stations to observe air temperature can be specified as follows:

```
instance OBSERVERS WeatherStation Temperature
    AmountOfAir where
        express (Temperature (AmountOfAir heat moisture))
            weatherStation = Measure heat Celsius
```

Expressing temperature is kept trivially simple here, using the numeric heat value as the number for the Celsius degree value. Of course, any other and more appropriate semantic datum can be implemented in this axiom.

6.5 Testing the Ontology

If an ontology is expressed in a programming language like Haskell, it needs to come with at least one model, in order for the code to be executable. The benefit of this is to make the theory testable through its model. Our experience has been that this is enormously beneficial for ontology engineering, as it reveals missing pieces, inconsistencies, and errors immediately, incrementally, and without switching development environments [42]. Constructing individuals of the categories under study (the above muensterAir was an example) and executing the operations reveals errors in axioms and provides the satisfaction (well known from programming) of seeing a solution "run" - though this is, of course, never a proof of correctness or adequacy.

7 Conclusions

This paper has presented a functional ontology of observation and measurement with the purpose to formalize and ground the semantics of observations. The ontology has the unique characteristics of

1. being functional in the sense of specifying observation as an information process independently of the technology involved;
2. generalizing over human and technical sensor observations;
3. distinguishing observations of endurant qualities from those of perdurant qualities;
4. specifying in what sense a sensor is a process, namely as playing the role of mapping stimuli to observations;
5. interpreting OGC's "phenomenon" as a quality of an endurant or perdurant;
6. tying OGC's "feature of interest" to physical endurants;
7. distinguishing internal (to the observer) and external bearers of qualities;
8. giving a formal account of qualia as quality constructor functions applied to endurants or perdurants;
9. supporting recursion in observation processes, thereby supporting sensor fusion;
10. remaining neutral with respect to the field vs. object distinction by keeping both views compatible with the ontology;
11. giving the ontology the form of a simulation, allowing for testing.

The ontology models qualities as data constructor functions, qualia (or internal sensor signals) as constructors applied to particulars, and observers as a role played by sensors, humans, or animals (to measure a quality). The observing behavior is abstracted into a type class that links observable entity and quality types to observer types.

At its current stage, the ontology represents only a first step toward a stronger ontological foundation for information sciences and technologies dealing with observations, particularly those with a spatio-temporal reference. It is now being extended along the following lines:

- instantiating the observer role for sensor systems and networks;
- treating sensor dust as a case where the individual observers are not positioned;
- capturing semantic datums (resulting from calibration) for the conversion of a detected signal to an observation value;
- specifying semantic datum transformations;
- generalizing the notion of observation to include affordances (as observed action possibilities);
- specifying the resolution of observations in space, time, and theme, based on the granularity of the sensed endurants and perdurants;
- modeling qualia and internal sensor signals as convolutions over space and time;
- modeling scale transformations;
- adding trust and reputation measures;
- extending the ontology by actuators, to simulate sensorimotor loops.

A translation of the resulting equational axioms into an ontology encoding language like OWL is straightforward and can be automated, but only warranted once the ontology is more complete and specific reasoning requirements (say, for sensor data fusion) have been identified.

Acknowledgments. Many discussions with Florian Probst, Krzysztof Janowicz, Simon Scheider, Sven Schade, Christoph Stasch, Arne Bröring, and Andrew Frank have shaped the ideas presented. The work on sensor ontology was initiated at the Research Workshop of our Institute with the National Institute for Space Research (INPE), Brazil, (see http://geochange.info), which has been supported by the German Research Foundation (DFG), project no. 444 BRA 121/2/09 and by The State of Sao Paolo Research (FAPESP), project no. 2008/11604-6. Additional insights resulted from our work in the International Research Training Group (DFG GRK 1498) on Semantic Integration of Geospatial Information.

References

1. Borgo, S., Masolo, C.: Foundational Choices in DOLCE. In: Staab, S., Studer, R. (eds.) Handbook on Ontologies, 2nd edn., pp. 361–382. Springer, Heidelberg (2009)
2. O'Sullivan, B., Stewart, D., Goerzen, J.: Real World Haskell. O'Reilly Media, Sebastopol (2008)
3. Goodchild, M.F.: Citizens as Sensors: the World of Volunteered Geography. GeoJournal 69, 211–221 (2007)
4. Broering, A., Janowicz, K., Stasch, C., Kuhn, W.: Semantic Challenges for Sensor Plug & Play. In: 9th International Symposium on Web & Wireless Geographical Information Systems (W2GIS 2009). LNCS. Springer, Heidelberg (2009) (in press)
5. Stasch, C., Janowicz, K., Bröring, A., Reis, I., Kuhn, W.: A Stimulus-Centric Algebraic Approach to Sensors and Observations. In: Trigoni, N., Markham, A., Nawaz, S. (eds.) GSN 2009. LNCS, vol. 5659, pp. 169–179. Springer, Heidelberg (2009)
6. Jones, R.S.: Physics as Metaphor (1982)
7. Stevens, S.S.: On the Theory of Measurement. Science 103, 677–680 (1946)
8. Heuvelink, G.: A probabilistic framework for representing and simulating uncertain environmental variables. IJGIS 21, 497–513 (2007)
9. Boumans, M.: Measurement Outside the Laboratory. Philosophy of Science 72, 850–863 (2005)
10. Gruber, T.R., Olsen, G.R.: An Ontology for Engineering Mathematics. In: Doyle, J., Torasso, P., Sandewall, E. (eds.) Fourth International Conference on Principles of Knowledge Representation and Reasoning. Morgan Kaufmann, Gustav Stresemann Institut, Bonn (1994)
11. McGhee, J., Henderson, I.A., Sydenham, P.H.: Sensor science: essentials for instrumentation and measurement technology. Measurement 25, 89–113 (1999)
12. Kim, H.M., Sengupta, A., Fox, M.S., Dalkilic, M.: A Measurement Ontology Generalizable for Emerging Domain Applications on the Semantic Web (2006)
13. Barnaghi, P., Meissner, S., Presser, M., Moessner, K.: Sense and Sens'ability: Semantic Data Modelling for Sensor Networks. In: Cunningham, P., Cunningham, M. (eds.) ICT-Mobile-Summit (2009)
14. Cox, S.: OGC Implementation Specification 07-022r1: Observations and Measurements. Open Geospatial Consortium (2007)
15. Chrisman, N.: Exploring Geographical Information Systems (2001)
16. Kuhn, W.: Semantic Reference Systems. International Journal of Geographic Information Science (Guest Editorial) 17, 405–409 (2003)
17. Kuhn, W., Raubal, M.: Implementing Semantic Reference Systems. In: Gould, M., Laurini, R., Coulondre, S. (eds.) AGILE 2003 - 6th AGILE Conference on Geographic Information Science, pp. 63–72. Presses Polytechniques et Universitaires Romandes, Lyon (2003)

18. Probst, F.: Observations, measurements and semantic reference spaces. Applied Ontology 3, 63–89 (2008)
19. Probst, F., Espeter, M.: Spatial Dimensionality as Classification Criterion for Qualities. In: International Conference on Formal Ontology in Information Systems (FOIS). IOS Press, Baltimore (2006)
20. Masolo, C., Borgo, S.: Qualities in Formal Ontology. In: Workshop on Foundational Aspects of Ontologies (FOnt 2005), Koblenz, Germany (2005)
21. Gärdenfors, P.: Conceptual Spaces - The Geometry of Thought. The MIT Press, Cambridge (2000)
22. Schade, S.: Ontology-Driven Translation of Geospatial Data. PhD thesis, Insitute for Geoinformatics (ifgi). Westfälische Wilhelms-Universität Münster, Muenster, 153 (2009)
23. Scheider, S., Janowicz, K., Kuhn, W.: Grounding Geographic Categories in the Meaningful Environment. In: Conference on Spatial Information Theory (COSIT). LNCS. Springer, Heidelberg (2009) (in press)
24. Gibson, J.: The Ecological Approach to Visual Perception. Houghton Mifflin Company, Boston (1979)
25. Liu, Y., Goodchild, M.F., Guo, Q., Tian, Y., Wu, L.: Towards a General Field model and its order in GIS. IJGIS 22, 623–643 (2008)
26. Couclelis, H.: Ontology, Epistemology, Teleology: Triangulating Geographic Information Science. In: Navratil, G. (ed.) Research Trends in Geographic Information Science, pp. 3–16. Springer, Heidelberg (2009) (in press)
27. Frank, A.: Tiers of Ontology and Consistency Constraints in Geographical Information Systems. International Journal of Geographical Information Science (IJGIS) 15, 667–678 (2001)
28. Buyong, T.B., Frank, A.U., Kuhn, W.: A Conceptual Model of Measurement-Based Multipurpose Cadastral Systems. URISA Journal 3, 35–49 (1991)
29. Leung, Y., Ma, J.-H., Goodchild, M.F.: A general framework for error analysis in measurement-based GIS I: the basic measurement error model and related concepts. J. Geogr. Systems 6, 325–354 (2004)
30. Buyong, T.B., Kuhn, W.: Local Adjustment for Measurement-Based Cadastral Systems. Journal of Surveying Engineering and Land Information Systems 52, 25–33 (1992)
31. Kuhn, W.: Editing Spatial Relations. In: Brassel, K., Kishimoto, H. (eds.) 4th International Symposium on Spatial Data Handling (SDH 1990), vol. 1, pp. 423–432. IGU, Zurich (1990)
32. Compton, M., Henson, C., Lefort, L., Neuhaus, H.: A Survey of the Semantic Specification of Sensors. In: 2nd International Workshop on Semantic Sensor Networks. A workshop of the 8th International Semantic Web Conference (ISWC 2009), Washington DC, October 25-29 (2009) (in press)
33. Sheth, A., Henson, C., Sahoo, S.: Semantic Sensor Web. IEEE Internet Computing 12, 78–83 (2008)
34. Probst, F.: An Ontological Analysis of Observations and Measurements. In: Raubal, M., Miller, H.J., Frank, A.U., Goodchild, M.F. (eds.) GIScience 2006. LNCS, vol. 4197, pp. 304–320. Springer, Heidelberg (2006)
35. Neuhaus, H., Compton, M.: The Semantic Sensor Network Ontology. In: AGILE Workshop on Challenges in Geospatial Data Harmonisation, Hannover, Germany (2009)
36. Webster, J.G.E.: Measurement, Instrumentation, and Sensors Handbook. CRC, Boca Raton (1999)

37. Botts, M., Robin, A.: OpenGIS® Sensor Model Language (SensorML) Implementation Specification. In: OpenGIS® Implementation Specification, OGC® 07-000, Open Geospatial Consortium, OGC (2007)
38. Percivall, G.: OGC Reference Model. OpenGIS® Implementation Specification (version 2.0), OGC 08-062r4. Open Geospatial Consortium, OGC (2008)
39. Juerrens, E.H., Broering, A., Jirka, S.: A Human Sensor Web for Water Availability Monitoring. In: OneSpace 2009 - 2nd International Workshop on Blending Physical and Digital Spaces on the Internet, Berlin, Germany (2009)
40. Ehrig, H., Mahr, B.: Fundamentals of Algebraic Specification. Springer, Heidelberg (1985)
41. Goguen, J.: Types as Theories. In: Reed, G.M., Roscoe, A.W., Wachter, R.F. (eds.) Topology and Category Theory in Computer Science, pp. 357–390. Oxford University Press, Oxford (1991)
42. Frank, A.U., Kuhn, W.: A specification language for interoperable GIS. In: Goodchild, M.F., Egenhofer, M.J., Fegeas, R., Kottman, C.A. (eds.) Interoperating Geographic Information Systems (Proceedings of Interop 1997), pp. 123–132. Kluwer, Norwell MA (1999)

The Case for Grounding Databases

Simon Scheider

Institute for Geoinformatics, Westfälische Wilhelms-Universität Münster,
Weseler Strasse 253, D-48151 Münster, Germany
simon.scheider@uni-muenster.de

Abstract. What is the *intended interpretation* of a geospatial database in terms of reproducible experiences? How should places on a digital globe be interpreted on the earth surface? And how can their spatial relations be reconstructed? How should road network databases be interpreted in terms of observable traffic infrastructure? And how is data about waterways and their depths to be interpreted in terms of observable water bodies? In this paper, I argue that successful information retrieval and querying of data in context-free environments requires that *such data interpretations need to be effectively coordinated*. I give four arguments why the current approaches to semantic engineering fail as methods in that respect, and why a 'grounding' approach to describe their semantics is necessary.

1   Introduction

In the early days of database research, issues related to database semantics played a prominent role in 'conceptual database design'. This began to change at the beginning of the 90's, and nowadays those issues do not appear to be part of mainstream database research [1]. The problem of data semantics regained attention in research areas concerned with databases in *context-poor communication environments*. In 'information system integration', it was recognized since the 90's that *semantic heterogeneity* is a major problem in order to handle arbitrary information requests, as users and providers need to mediate their understandings [2]. Because a special case of such a communication environment is the web, the idea of the 'semantic web' was launched later in 2000 [3].

From the very beginning of research in semantics, ontologies, formal theories of the commonsense world first introduced in artificial intelligence [4], were proposed as a major tool for meaning description. At the same time, the sufficiency of *formal symbol systems* to construct intelligent agents was debated in artificial intelligence [5]. One argument was the *'symbol grounding problem'* [6], the problem that declarative semantics expressed in terms of formal symbols gives rise to an infinite regress. Since then, ontologies have been successfully used for describing geospatial data. But even though the available ontological tools have matured with respect to expressiveness and computability, fundamental debates about their adequacy for a semantic strategy never stopped [7]. It was recognized quite early in spatial reasoning that the main challenge to represent spatial concepts lies in the need for different and context dependent world views [8]. One

K. Janowicz, M. Raubal, and S. Levashkin (Eds.): GeoS 2009, LNCS 5892, pp. 44–62, 2009.

progression was towards more collaborative approaches based on user-generated 'folksonomies' [9]. Another was to use semantic similarity [10], schema mapping and integration tools.

Meanwhile, the problem has remained pretty much the same: heterogeneous information communities need data from diverse external sources in order to produce new products and services, but these products were mostly *not intended by the data suppliers*. Therefore the discovery, retrieval, integration and query of such information requires from the user that he be able to reconstruct its *intended meaning*.

As Kuhn argued in [11], despite the notoriously difficult philosophical questions involved, *semantic interoperability* can be seen as an *engineering problem*. Solutions should be based on minimal and sound methodological assumptions and a clarification of the scope. In this paper, I argue that *effective semantic methods need to coordinate human interpretations* (Sect. 2). I discuss the fitness of present methodological approaches in this regard by reviewing their *underlying assumptions*: Determinism of scientific thought in natural science, established usage of natural language, and precision in declarative semantics of formal languages (Sect. 3). I make the case for a *'grounding' method* (Sect. 4) through *four arguments*, each of them showing how these assumptions are challenged by the *indeterminacy of human interpretation*.

2   Database Queries Need to Relate Human Interpretations

Borgida's and Mylopoulos' revised idea about data semantics [1] will serve as a recurrent theme in this paper. They propose to see *data as a 'model' whose purpose is to answer questions* about the subject:

> "Consider, for example, the case of a geopolitical globe as a model of the Earth.[...] the (informal) questions to be asked have to do with the existence and (relative) position of features on the Earth's surface, [...]. The questions about the model are answered by direct observation of the model by a human, aided perhaps by a string/ruler/compass [...]. The mapping of questions and answers is based on the scale reduction of the Earth's spherical surface to that of the globe." [1]

There are two far-reaching lessons hiding behind this simple example:

Data can be interpreted by humans: First, this case suggests that a data set should *have meaning in the same sense as natural language has*. Effectively, there even should be a *natural language analogue* to a data set that humans can *understand* and *practically interpret* in terms of their own situation. Otherwise the data query could not produce meaningful answers in natural language. This notion of meaning essentially involves that the data set should be *interpretable in terms of human experience and human thought*. In our example, the computation of the relative location of Paris and San

Fig. 1. A database query model

Francisco (Fig. 1) requires both. This seems to be a rather trivial fact, but it is not: main stream computer science since the times of Ted Codd's invention of the relational database model has sticked to another view. Semantics was factored out during run time of a database: Database semantics was managed by the operational environment of a database system, i.e. its database administrator and application programs, and "If you wanted to know what the data really meant, you'd have to talk to the administrator"[1]. Database queries — following the 'physical symbol systems hypothesis' [12] — were considered to be formal manipulation of symbols without any necessary further meaning. But it is this very precondition of *human interpretability of databases*, which seems to make the implementation of an intelligent digital program, like the question-answering *'Turing Machine'*, a mere fiction[1]

Interpretations can be related: Second, the mapping between questions and answers in the data query is obviously based on a *relation between different interpretations, i.e. physical observations*: the relation of the earth surface to the globe model. Note that what is being related here are observations. The person posing a question must be able to interpret an answer in terms of a place on the observed earth surface. But the answer was derived by another person from its observation of the globe surface. *This relation is exactly how — through a chain of interpretations — human-interpretable questions are linked to human-interpretable answers.* And it is the reason why the database query actually works, that is, why it is able to deliver meaningful answers to humans.

What exactly is meant here with the notion *'interpretation'*? Obviously something very similar to von Glasersfeld's concept in [13]: An interpretation is a *decoding of symbols in terms of thoughts and experiences*. More specifically, it is

---

[1] John Searle argues in [5] convincingly that symbol manipulation is not *sufficient* for understanding a natural language text, because *understanding* involves *meaning*, i.e. the possession of intentionality: the possession of mental states, like e.g. beliefs, desires and intentions, that are *directed at states of affairs in the world*.

a *mind specific activity of a human interpreter S* taking an *immediately experienced object X (the semiotic object, e.g. a symbol)* and producing *a not immediately experienced result Y*, which is not part of X. X could for example be the question that person S tries to answer by interpreting it on the experienced globe Y. If X — like in our case — was created by another person A, the 'author', then A's own individual interpretation similarly may have produced M, the *'intended meaning'* of X, which is the earth surface experienced by person A (see Fig. 2).

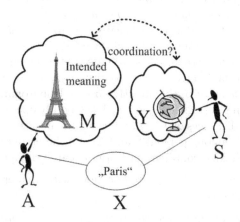

Fig. 2. How to coordinate interpretations is the real challenge for a successful database query

The view that the semantics of a *formal vocabulary also* requires interpretation in terms of thoughts and experiences (sometimes called 'conceptualization' in order to stress their interdependence), has been common from the early days of logic based research in artificial intelligence[2]. This view seems to be agreeable among ontologists (see e.g. Guarino [15]), because it contradicts few philosophical world views (apart from, perhaps, a platonistic one). I will use it as a starting point for my discussion.

But I would like to stress two aspects of potential disagreement. First, thoughts and their interpretative association are *artifacts of the individual human mind* [13]. The two interpretation activities in our example produce results in two different heads, so *Y never happens to be the same as the intended meaning M*. This fact about interpretation therefore stands in sharp contrast to any 'conduit metaphor' of language [16], or any naive idea of a 'universal concept'. Second, whether for each symbol there is a stable interpretation, or whether more than one thought

---

[2] "In making our definition [of semantics], we assume the perspective of the observer [...]. We have a set of sentences and a conceptualization of the world, and we associate the symbols used in the sentences with the objects, functions, and relations in our conceptualization" ([14], Chap. 2.3).

can encode one symbol, that is, the *semantic rules*, seem to be a private affair in the first place. The rigorous association implied by formal semantics is therefore not necessarily the case in human interpretation. If we strictly define a symbol in terms of thoughts or experiences, we arrive at a close association of thoughts. If we loosely relate a symbol to our web of beliefs, there can be much more variation in terms of facts. These obvious degrees of freedom will be our main concern in this paper.

Obviously, semantic interpretation poses a challenge to every database query. However, as we see in our example, *the private interpretations are often successfully coordinated in such a way as to give meaningful answers to questions*. I will suggest in this paper that this is so because *questions and answers are both grounded in common observable operations*: the answer of S is grounded in geodetic measurements, which are themselves grounded in the human observation of the environment, and this indirectly observed environment of S happens to be closely coordinated, *on this level*, with the directly observed environment of A. The two interpretations are on a certain level on which the bodily operations — for example the measuring of distances on the globe by S and on the earth surface by A — are *mutually referable* among persons S and A, and therefore related.

This is the implicit idea contained in Borgida's and Mylopoulos' example:

Claim. Database queries need to relate questions with answers by coordinating their human interpretations.

Once this problem of relating interpretations is solved, the problem of semantic heterogeneity is solved, too, as any query posted to a data set could be given an answer related to the question's *intended meaning*. But this turns out to be the real problem, because language interpretation is a largely indeterminate process, and coordinating those processes is a challenging task.

3   The Indeterminacy of Language Interpretation

In this section I will discuss evidence and arguments for the view that semantic interpretation of language in general is a vague, underconstrained and indeterminate process. The indeterminacy of this process challenges the *existing approaches to semantic engineering*, because it undermines their assumptions: Determinism of scientific thought in natural science, established usage of natural language, and precision in declarative semantics of formal languages.

3.1   The Argument of Indeterminacy of Empirical Theories

The first argument is one of epistemology, that is the justification of knowledge. If a web of beliefs[3] was shared among interpreters, the coordination task could

---

[3] In epistemic logic, knowledge is considered as true belief, i.e. the special case of believing in true propositions. But knowledge is a much debated notion, and truth even more. In this section I am talking about webs of human beliefs in that general sense, i.e. about interrelated proposition that are *accepted* as being true, regardless of being also called 'knowledge' (thanks to one reviewer for his hint).

be reduced to *negotiating a common vocabulary* for the notions involved in those beliefs. The argument has much to do with the methodological implications of *philosophical realism* (also called *metaphysical objectivism*) [17].

If we assume that our web of beliefs can be harmonized by approximating the 'real world' through observation and scientific reasoning, then this could be a way of coordinating interpretations. If the real world e.g. consists of discrete objects and their relations, then agreement on interpretation is just a function of the correctness and completeness of representing those relations in thought and language. Perhaps, one could admit that only experts really have access to these correct descriptions, like Putnam did [18], and thus require ontologies to be constructed solely by experts.

Realism is in fact quite often used as a justification for a certain methodology of ontology engineering. One example is Barry Smith's proposal [19] to replace the notion of a *concept* as the subject matter of ontologies by "the universals and particulars which exist in reality". The idea is that the ambiguity of ontological terms as well as the existence of different ontological views on a subject matter, which seem deeply entangled with the imperfectness of human perception, will disappear once ontology engineering is "devoted precisely to the representation of entities as they exist in reality" [19], and not to mere linguistic or cognitive artifacts. In this view, people *hold different views on reality due to human misconception*, but once they strive for a better (less subjective and more natural-science compatible) view, their conceptions will converge. A comparable philosophical view stands behind many attempts to formalize, once and for all, a so called *universal ontology*, in which agreed human knowledge is encoded.

It is well known that scientific realism is confronted with a large list of powerful criticisms[4]. Please note that I will not engage in any metaphysical discussions about realism. I personally think that a 'basic realism', e.g. in Lakoff's sense [21], is beyond question. But I would like to demonstrate the *practical methodological problems* that one will encounter when following a realist approach in order to harness the effects of the indeterminacy of language interpretation.

Empirical Underdetermination. One of the most powerful arguments against philosophical realism is based on the so called *empirical underdetermination hypothesis* [22]:

Claim. For every empirical theory T there is an *empirically equivalent* theory T' which is incompatible with T.

The term 'empirical equivalence' is explained by Richard Boyd [17] in the following way:

> "Call two theories empirically equivalent just in case exactly the same conclusions about observable phenomena can be deduced from each. Let T be any theory which posits unobservable phenomena. There will always be infinitely many theories which are empirically equivalent to T but

---

[4] See [17] and [20] for an extensive discussion.

which are such that each differs from T, and from all the rest, in what it says about unobservable phenomena (for formalized theories, this is an elementary theorem of mathematical logic)." [17]

The argument of empirical underdetermination does not force one to assume that there is no objective world outside of human thought, or that progress in scientific reasoning is not possible. But it leaves no reason to think that sophisticated conceptualizations of the real world will *automatically and asymptotically approach a unique state of agreement*.

Duhem, Quine and Knowledge Discovery The power of the underdetermination argument can best be illustrated by an example. W.V.O. Quine [23] gave one: According to physical theory T, we live in a universe that can be described by an infinite 3-dimensional Euclidean space. However, according to theory T', we live inside of a 3-dimensional ball, and the closer an object approaches its surface, the smaller it gets. Empirically, by using observation or measurement, it is impossible to distinguish between the two worlds that are described by T and T', because the measurement units will shrink in exactly the same way as the objects do. Nevertheless, the theories are incompatible because in T', an unobservable center of the universe exists, but not in T.

The idea of empirical underdetermination is a radicalization of the famous *Duhem-Quine thesis*. Empirical contradiction (often called *falsification*) can — according to Duhem [24] — never be accomplished for an isolated hypothesis. The argument is that *observable implications* never exist for a *single hypothesis*, but only for a conjunction with auxiliary premises. Falsification therefore only implies that any assumption, including the hypothesis, could be false, but which one is left undecided. The argument can be extended to *whole theories* as sets of sentences, their inference rules and derivable facts. As Quine put it: "Any statement can be held true come what may, if we make drastic enough adjustments elsewhere in the system"[5]. He devises an alternative conceptual view on empirical knowledge, which might be called the *'undetermined fabric of knowledge'*:

> "The totality of our so-called knowledge or beliefs, from the most casual matters of geography and history to the profoundest laws of atomic physics or even pure mathematics and logic, is a *man-made fabric* which *impinges on experience only along the edges*. A conflict with experience at the periphery occasions readjustments in the interior of the field.[...] Re-evaluation of some statements entails re-evaluation of other, because of their logical interconnections — the logical laws being in turn simply certain further statement of the system [...]. But the total field is so undetermined by its boundary conditions, experience, that there is much

---

[5] See [25]. Quine's critique is directed towards what he calls *reductionistic verification* in empiricist thought (popularized e.g. by Locke, Hume as well as Carnap), namely the view that every statement of a theory is translatable into a statement about immediate experience and thus can be evaluated in isolation.

latitude of choice as to what statements to re-evaluate in the light of any single contrary experience." ([25], emphasis mine)

This so called *holistic* argument is not really in danger by saying that in concrete cases, researchers usually can agree on which parts of a theory to retain by referring to preferential rules and heuristics. The problem can be illustrated by realizing that is has practical analogues in the field of *knowledge discovery* and *statistics*: In order to learn a curve from observations e.g., it is always possible to choose among a set of linear and non-linear regression rules. But this choice is often not clearly decidable from empirical evidence: it needs a *theoretic bias*. There is lots of heuristic strategies to select from [26], e.g. simplicity and *Ockham's razor*. But it is not the case that this rule — or any other theoretic bias — is most likely to be successful a-priori [27].

Objective Ontology and Natural Science. Taking these arguments into consideration, it seems naive to assume that even in the most accurate natural sciences, e.g. physics, there will emerge one *objective ontology* which contains all abstract concepts of the discipline. This of course does not mean that it is impossible to build a single meta-theory in physics. It just means that given the empirical facts, different ontological views must always be expected.

Take e.g. the abstract concept of *matter*. In contrast to *mass* or *energy*, which are closely tied to observation procedures, matter is nowadays banned from any physics textbook, even though it has always been a core concept from the outset, for example in Aristotle's metaphysics [28]. Since then it has been used throughout physics in a bewildering variety of contexts[6].

It is important to note that even though quantitative physics is a very stable success story of modern science, its *ontological interpretation* continues to give puzzles and seems to be far from reaching any agreement. The different existing interpretations of the *mass-energy equivalence* may serve as one example [30]. As another example, the ontological analysis of the experiential findings in quantum mechanics of light, as discussed by Bohr, Schrödinger and Einstein, resulted in the wave-particle dualism, which eventually made two contradictory interpretations permanent.

Can we thus conclude that at least that part of physics which is less ontologically abstract and more closely related to measurement is not affected by the indeterminacy argument? The indeterminacy argument was applied to physics by P.W. Bridgman ([31], Chap. 8). Bridgman argues that the *accuracy limitations of physical measurement*, especially in wave mechanics, seem to have opened the door for a flood of what he calls 'possible' theories. Obviously, many macro scale concepts of ordinary experience, like 'wave', 'particle' or 'probability' seem to become metaphors at the scale of quantum mechanics, because they are not applicable in a strict sense ([31], Chap. 9). In consequence, there is lack of knowing which parts of the mathematical theory shall be interpreted as real and which parts as mere formal artifacts.

---

[6] "It is fair to say that in contemporary physics, there is no broad consensus as to an exact definition of matter"[29].

There are of course many more obvious fields of natural science than physics to demonstrate the effects of empirical underdetermination. A particularly relevant and frequently discussed example is taxonomies. As Lakoff ([21], Chap. 12) elaborates in detail, the scientific history of *biological taxonomies of species*, a classical example of a scientific ontology, seems to contradict the objectivist assumptions: currently, there remain three incompatible views on taxonomy. Why should one claim that there must be a *predisposed, unique and correct* way of categorizing species? The troubles associated with such a view amount to an absurd level when one considers the case of the duck billed platypus, as described in Eco ([32], Chap. 4). Because the duck billed platypus has got mammary glands as well as eggs, it caused considerable confusion among taxonomists of the 19th century. The confusion ended up in a *conventional taxonomic revision* and a new category 'monotrema'. Interestingly, Eco points out that this was the result of *80 years of scientific debating*, and that during that debate, the *encyclopedic characteristics*, like being a mammal or not, were fully negotiable, whereas the *directly observable* characteristics of the animal, like its tail and duck bill, were *indelible*.

Observability. It has often been brought up against the empirical underdetermination argument, that it draws on the distinction between *'observable'* and *'unobservable'* sentences in a theory, and that this distinction cannot sharply be made, or that it is theory dependent itself. First of all, observability seems to be so crucial a notion that I do not see how to do without it even in a realist setting. It is necessary to have an *operational understanding* of a theory in order to justify it with empirical data. Furthermore, as Boyd [20] demonstrates, such a distinction must not be sharp, it could be a continuous transition from observable to unobservable, and can be sharpened if needed.

In fact, the notion of observability can be made precise in the way Quine did, for example in [33]: Quine's argument is that natural language sentences vary in their semantic indeterminacies. There are certain *occasion sentences*, utterable only on the occasion, with relatively low indeterminacy and high observability, like 'it's raining' or 'it's a rabbit'. These sentences are called *observation sentences*: "An observation sentence is an occasion sentence that the speaker will *consistently consent to* when his sensory receptors are stimulated in certain ways [...]" ([33], emphasis mine). This does not necessarily mean that the observation terms (categories) in these sentences must be 'coextensive' or pointing to exactly the same things[7]. It just means that the *individual interpretations are coordinated* in a way to allow successful communication[8], and this is — as we saw in Sect. 2 — exactly what is required from semantics.

---

[7] Following one of his famous examples, 'Gavagai' could denote a rabbit as well as mere "stages, of brief temporal segments, of rabbits" ([34], page 51). The two meanings are not distinguishable by pointing at rabbits.

[8] A 'successful' communication process could be considered as guided by Grice's *cooperative principle* [35], so that communicative acts follow mutually accepted expectations.

This effectively means that there must be consensus about names for bodies and body parts. According to Quine ([34], Chap. 3), the agreement on names for bodies like 'Mama' can be based on an observable action such as ostension, given the situation is simultaneously observed and the viewpoints of a language teacher and a learner are enough alike. In the same manner, the correct word usage is inculcated in the individual child of a language community by social training on the occasion, that is by the child's disposition to respond observably to socially observable situations, and the adults disposition to reward or punish its utterances ([34], Chaps. 1 and 3). Observation sentences are *the entrance gate to language*, because they can be easily learned directly by ostension *without reference to memory or theory* ([36], §11). In this way, observation terms actually spread far beyond the concretely observable situation and consistently recur in different theories[9].

The important thing to notice here is that any *critique towards observability* which relies on its *theory-dependence* or on its *social dependence* misses the point, because

> "The problem of relating theory to sensory stimulation may now be put less forbiddingly as that of theory formulations to observation sentences. In this way we have a head-start in that we recognize the observation sentences to be theory laden. What this means is that terms embedded in observation sentences recur in the theory formulations" [33].

Conclusion. Because the human 'fabric' of empirical knowledge is largely underconstrained by observation, different and incompatible conceptualizations of reality have to be expected at any time. The practical consequence is: abstract, less observable concepts in a given ontology cannot be expected to have an equivalent counterpart in a second ontology, which makes translations and ontology mappings in part a bold venture. Furthermore, any commitment to realism will by no means prevent this, and therefore will not be a solution to the problem of semantic heterogeneity.

This means that the semantic engineer confronted with semantic heterogeneity problems basically is left without the 'God's eye view', but not without *reference to collective experience*, and not without 'truth' or 'rationality', as Putnam argues[10].

3.2   The Argument of Indeterminacy of Natural Language Use

One could argue that because natural language is practically successful in conversations, in a way that seems to perfectly constrain the intended meanings of

---

[9] Quine points out that in scientific discourse, observation sentences are the 'common ground' to fall back on [36].

[10] "'Truth', in an internalist view, is some sort of idealized rational acceptability, some sort of ideal coherence of our beliefs with each other and with our experiences, as those experiences are themselves represented in our belief system — and not correspondence with mind-independent or discourse-independent 'states of affairs'" ([37], Chap. 3, page 49-50).

words and sentences, why not solve the problem of semantic heterogeneity by sticking to natural language descriptions? There are definitely parts of natural languages whose interpretation is strongly constrained by the language community. As John Searle [38] points out convincingly, language is a constitutive part of *constructed social reality*, especially in order to establish *'objective' institutional facts*, like e.g. the supreme court making a decision, or the assignment of a status like 'money' to a piece of paper. It is important to notice that there are obviously *subsets of natural language which already come with coordinated interpretations*, because they are results of the collaborative construction of social reality. I would count Searle's *institutional facts*, but also Quine's *observation sentences* here, which were discussed in the last section.

One may think that a determined use and interpretation of language is a social fact itself. But is the interpretation of an arbitrary word fixed in the same way as the interpretation of a certain piece of paper as being money? It turns out that an important source of indeterminacy of natural language interpretation is its degrees of freedom in usage. Linguists have frequently pointed out that natural language usage must be a *creative and open process*, which makes use of these degrees of freedom to account for the fact that an unlimited variety of meanings must be expressible using only a very limited human lexicon [39]. One observable effect of this is the frequent *re-use or re-interpretation* of lexical symbols, as well as their *metaphorical use*.

Linguists have discovered many sources for semantic indeterminacy of language, e.g. *graded structures* and *prototype effects* of category words (compare the discussion in Chaps. 4, 5 and 6 of Lakoff's book [21]), like in the case of the word 'mother'[11]. Sometimes categories are even used *metaphorically*, and it is unforeseeable which aspects of a prototypical meaning motivate those category variations under the umbrella of a single word, the border case of which are totally 'unrelated homonyms' (see Lakoff's example of the Japanese 'hon', Chap. 6 in [21]). Lakoff also observed (see [21] Chap. 8), that cognitive models underlying the semantics of language sometimes *need to be inconsistent*, because language can be used to talk about itself in the same sentence. For example, *negation* can mean to deny the truth of either a whole cognitive model, or of its fore- or background conditions. It is only apparent from the communication context in which way negation must be interpreted.

It is this *context dependence* of meaning in natural language that makes it flexible as a tool for human communication, but does not render it an appropriate tool for constraining semantic interpretations in a context-free setting. This is I guess what is really meant by saying that human language is 'imprecise', whereas formal logic is 'precise'. Negation in logic has one and only one interpretation, because its syntactic and semantic rules are explicit. So it seems that formal languages are *necessary* to restrict interpretations effectively in

---

[11] The genetic, the nurturance, the marital and the genealogical aspects of motherhood are all fulfilled by a prototypical mother, but there are non-prototypical usages of the word, e.g. 'genetic' mother, stepmother and so forth, that have nothing in common, but seem to agree only with arbitrarily selected aspects of this list.

context-free communication environments. But does this mean that formal theories are also *sufficient* for coordinating interpretations?

3.3   The Argument of Unintended Semantic Domains

One might think that ambiguity of language interpretation may be a problem only for natural languages. But as I will show in the next two subsections, the problem exists for *natural* as well as for *formal languages*. Therefore semantic heterogeneity problems can always be found in both, and, as a surprising consequence, *formalisms alone turn out to be an inadequate means to solve these problems*. It becomes apparent that the symbol grounding problem [6] is a very essential constraint for all languages, which implies deep practical problems, because *meaning — understood in terms of semantic interpretation of symbols by humans —* is not conveyable in any language alone.

The first argument is concerned with the *identification* of what is called a *'domain of interpretation'* in model theoretic semantics, that is a set in terms of which the symbols of a formal language are being interpreted. In first-order logic (FOL), a *signature* is a set of constant, predicate, and function symbols. Together with the syntactic rules of FOL it gives rise to a language (the set of well-formed formulas, or sentences). Now, in 'model theoretic semantics', to interpret a signature, sometimes called *'a structure' of the signature*, means

1. to identify a *concrete set (called 'domain', e.g. D)*, and
2. to associate each *constant* symbol with an *element of D* and each *function/predicate symbol* with a *concrete n-ary relation/function in D*.

Tarski's model theoretic truth definition tells us when a given sentence is true in this structure, in which case we can talk of the structure as a *'model' of the sentence*. Model theory is almost exclusively about this second aspect of a structure, that is it is assumed that *D* together with its concrete relations is given, and we just look at those interpretations that preserve the asserted truth of, i.e. *satisfy*, certain sets of sentences, called theories.

But what exactly is meant with the first aspect? What does it mean that a *concrete set 'is given'*? The problem is that *meaning*, as I described it in Sect. 2, is an interpretation of symbols into a *very specific domain*, namely into the domain of our thoughts, experiences and mental operations. As we saw, this domain is in no way *'given'* in the sense that everyone has equal access to it. Among model theorists, it is implicitly assumed that everyone knows what one is talking about when talking about domains of interpretation. *But this turns out to be the real problem of semantics.*

There is a method of making us aware of the personal mental domains we use when we interpret signs. It makes use of the idea of *analogical representations*, which are being discussed in the seminal works of Sloman [40] for artificial intelligence and Palmer [41] for cognitive science. Sloman argues for the existence of analogical representations including truth-values and even valid inference procedures: "Discovering the truth-value requires the application of semantic interpretation procedures in investigating the world"[40]. The idea is that if we investigate the world around us *and* if we interpet a visual sign, we always *experience concrete relations* between parts of (what Sloman calls) a "configuration", and these relations can be used to make valid statements and even inferences. For example, look at Fig. 3: It is a visual configuration that represents 4 objects ordered with respect to their tallness. We are immediately able to recognize an order relation among the 4 objects: 'Taller than' is a fundamental mental operation we are used to apply to spatial objects. Note that this operation has logical properties, e.g. if a is taller than b and b is taller than c, a is also taller than c (transitivity). Note also that there is only *one way* to order the objects according to tallness, even though there is *not any sign in this picture explicitly describing the relation* (it is a pure mental construction).

Figure 4 is a representation of the tallness relation by another ordering operation, 'longness'. Note that this representation does not have widths anymore. It is *operationally poorer*, but also *less ambiguous*. As 'Longer than' is very similar to 'Taller than', this representation could also be called *'iconic'*. In Fig. 5, 'Taller than' is represented by 'Points to'. Here, almost every iconicity is lost (apart from the correspondent sequence of objects from left to right), but an *explicit sign* for the relation 'Taller than' is present, the arrow. Figure 6 is actually a *formal first-order theory* representing tallness. There is only *one mental operation left*, namely *function application* (I call it 'Takes'), which has to be constructed while reading the text. Every semantically relevant mental aspect was explicitly converted to an atomic or constructible sign: the 4 objects and

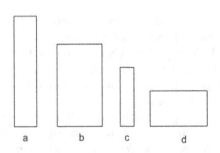

Fig. 3. *Analogical* representation of 'Taller than'

Fig. 4. *Analogical* representation of 'Taller than' by 'Longer than'

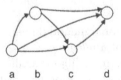

Fig. 5. *Analogical* representation of 'Taller than' by 'Points to'

$a \geq b$

$b \geq c$

$c \geq d$

$\forall x,y,z.\ (x \geq y \land y \geq z \rightarrow x \geq z)$

[function] Takes [arguments]

Fig. 6. *Fregean* representation of 'Taller than'

their immediate tallness-neighbors are written down into 4 facts. All other facts, e.g. that a is taller than d, can be deduced by applying the 5th fact, the transitivity axiom. As we see, all of these representations need mental operations, but to a different degree, which makes them more or less flexible to represent other mental operations. The most flexible one is of course 6, which Sloman calls a *Fregean representation*: "A Fregean system has the advantage that the structure (syntax) of the expressive medium need not constrain the variety of structures of configurations which can be represented [...]" [40].

I would like to allude to a point which seems to have been overlooked by both authors. The *power of constraining the possible sign interpretations — better: of reducing the potential mental domains of interpretation — increases dramatically from Fig. 6 to Figs. 3 and 4.* There are two ways of applying a mental order operation in 3 and just one way in 4. Because longness in Fig. 4 is iconic to tallness, Fig. 4 is even able to hint at the correct interpretation. But think about how many mental domains of ordering could be denoted by Fig. 6, *unless we already know* that it is supposed to represent only 'Tallness'? It could be interpreted in terms of every domain with a partial order, that is, *nearly our whole universe of thought*: natural numbers, real numbers, the incomes of citizens, the distances of planets from the earth, etc.

Claim. Fregean representations, which are used in formal languages, are inherently incapable of *indicating their domains of interpretation*, because they always allow for unintended domains. Thus they cannot be used as a method to coordinate semantic interpretations with respect to domains of thought and experience.

3.4 The Argument of Indistinguishability of Reference

Now that it is clear that formal (Fregean type) representations cannot indicate their semantic domains: are they still capable of fixing the *exact reference*, that is the intended correspondence of thoughts and symbols *inside of a given domain*?

The answer is *in general no*, and this is an important result of Hilary Putnam's theorem stated in *Reason, Truth and History* ([37], pages 217–218):

> "Let L be a language with predicates $F_1, F_2, ..., F_k$ (not necessarily monadic). Let I be an interpretation, in the sense of an assignment of an intension to every predicate. Then if I is nontrivial in the sense that at least one predicate has an extension which is neither empty nor universal in at least one possible world, there exists a second interpretation J which disagrees with I, but which makes the same sentences true in every possible world as I does."

Putnam basically says here that in model theoretic semantics, *truth of a sentence can always be maintained while its reference is changed*: "No view which only fixes the truth values of whole sentences can fix reference" [37]. If Putnam's result is correct, this means that there is always a second, different interpretation of a given theory in terms of a given domain, no matter how precise or detailed a theory is. No formal description will then be sufficient to determine reference within model theory. This is Putnam's main argument to reject model theory as a theory of meaning, and to propose that semantic reference is 'direct' — not 'indirectly' fixed via descriptions of properties, but *directly via acts of naming* (see also the discussion in Lakoff [21]).

The phenomenon is known in the ontology community under the name *'unintended models'*, mentioned already in Hayes' early account of ontologies in [4]: "Indeed, no formal operations, no matter how complex, can ever ensure that tokens denote any particular kinds of entity" [4], and also in Guarino's foundational paper [15].

But the considerable epistemic trouble with this insight is that there is *no way of fixing reference within a language*. Take for example the notion *reference* itself. Is it possible to fix reference by precisely describing it in a language? There have been actually many attempts like this, e.g. defining *x refers to y if and only if x bears R to y*, where R would be a relation that characterizes the reference relation. But such a theory would have to characterize R by another collection of sentences. By Putnam's theorem, these sentences will still allow many models for R, so there is inevitable indeterminacy again.

To see the practical consequence of Putnam's theorem, let us look at a well known example from measurement theory. As Suppes points out in [42], measurement scales are maps from some observable structure (qualities in some terminologies) to a set of symbols. Measurement theory describes formal properties for such a reference map, namely scale types. But scale types do not disambiguate scales themselves. Individual scales, e.g. the 'meter' length scale, *are not uniquely determined by their formal structure*. This is called the *uniqueness problem of measurement* [42]. And it is the reason why *measurement standards*, like units of measurement, are necessary in order to fix the degrees of freedom, and in order to interpret the scale in the intended way.

In the case of our *globe example* from Sect. 2, the reason why Putnam's devastating result does not affect the interpretations of the database query, is that those interpretations are — very similarly — related by a *geodetic datum* for a *spatial reference frame*, e.g. an ellipsoid. A geodetic datum for the positions on a Bessel ellipsoid consists for example of a named spot on the earth's surface like 'Rauenberg' near Berlin (Potsdam Datum), and a standard position and orientation for the ellipsoid. All of these have to be physically realized by appropriate observations and operations.

4   Grounding Data as a Way of Coordinating Its Semantic Interpretation

We have seen that the solution is neither contained in natural science, nor natural language descriptions, nor formal theories. I suggest that the only way of fixing and relating semantic interpretations is through a process called *'grounding'*. This process will involve reference to the human body and his 'Gestalt' perception of the environment, but also competences of collective naming and ostension. Looking at our example from Sect. 2, we suggest that the *competence of relating individual interpretations* seems to lie in a *collective competence of humans* that has at least three parts:

1. the competence of (mutually) referring to collective and reproducible sensory-motor experience,
2. the competence of establishing a common 'observation language' about it (a language of empirical facts in the sense of Quine's 'observation sentences')
3. and the competence of expressing ambiguous symbols in terms of these observational primitives.

The first and the second competence together e.g. enable the person with the globe and the person in Paris to refer to the same object, like e.g the 'Eiffel Tower', and to refer to orientation concepts like 'north of', as well as to measurement standards. The third competence enables the person with the globe to 'ground' complex reference frames, like e.g. an ellipsoid or the globe, and thus to interpret complex calculations as operations on the experienced earth surface. For example, two points on a longitudinal circle on the globe can be interpreted in terms of the 'north of' relation. This third competence involves exactly the idea of *'semantic reference systems'*, that were first described by Werner Kuhn [43].

We could say that the competence of humans to coordinate interpretations of a symbol is given by *operationalizing* that symbol, that is, to say what it means in terms of mental or physical operations. But this view should not be reduced to a naive operationalism, which tries to restrict the meaning of every word to concrete instances of physical operations, which are often subject to change and non-repeatability. From the discussion above, it should also be clear that there is no computable strategy for deciding about the correspondence of operations and symbols, since it would again run into the symbol grounding problem.

Our view involves the establishment of a primitive operational language about observation *as a social fact*, and it therefore needs the social re-construction of observation symbols by the act of pointing to something others can repeatably observe as well. It is this mutual act of focusing the attention to sensors and experiences in a group of people which gives rise to the coordination of interpretation in that group. Once such coordination is achieved for a small set of symbols, other symbols can be related to this language by saying what it means operationally in that language. It is just about what Quine called the 'edges of our web of beliefs which impinge on experience' or 'observation sentences', which are the basis for a method of grounding. Whether the web of knowledge as a whole is correspondent or not between humans is not in the focus and not required for this method. Also, it will not have to draw heavily on natural language with its multitude of usages and ambiguous notions, because it can introduce new symbols. And as the establishment of social facts needs clear and established methods of construction, symbol usage, and inference, a formal language is obligatory. But such formal theories do not determine interpretation, they are more an *inductive consequence* of a previously established way of using and interpreting the symbols.

I have given an example for such a formal observation theory based on Gibson's meaningful environment [44] and a generalized account of a 'sensor' in [45]. It can be used to define geographic data categories, like e.g. *water depth* and *road networks*, in terms of collectively observable primitives, like sensory-motor affordances, the surface layout of the environment, and its perceivable geometry. The theory is currently under development and is intended to be applied to diverse semantic domains of geographic information.

Acknowledgments. This work is funded by the Semantic Reference Systems II project granted by the German Research Foundation (DFG KU 1368/4-2).

References

1. Borgida, A., Mylopoulos, J.: Data semantics revisited. In: Bussler, C.J., Tannen, V., Fundulaki, I. (eds.) SWDB 2004. LNCS, vol. 3372, pp. 9–26. Springer, Heidelberg (2005)
2. Sheth, A.P.: Changing focus on interoperability in information systems: From system, syntax, structure to semantics. In: Goodchild, M.F., Egenhofer, M.J., Fegeas, R., Kottman, C.A. (eds.) Interoperating Geographic Information Systems, pp. 5–30. Kluwer, Norwell (1999)
3. Berners-Lee, T., Hendler, J., Lassila, O.: The semantic web - a new form of web content that is meaningful to computers will unleash a revolution of new possibilities. Scientific American Magazine (2001)
4. Hayes, P.J.: The second naive physics manifesto. In: Hobbs, J.R., Moore, R.C. (eds.) Formal theories of the commonsense world. Ablex series in artificial intelligence. Ablex Publ. Corp., Norwood (1985)
5. Searle, J.R.: Minds, brains, and programs. Behavioral and Brain Sciences 3(3), 417–457 (1980)
6. Harnad, S.: The symbol grounding problem. Physica D (42), 335–346 (1990)
7. Shirky, C.: Ontology is overrated,
   http://www.shirky.com/writings/ontology_overrated.html
8. Freksa, C., Barkowsky, T.: On the relation between spatial concepts and geographic objects. In: Burrough, P., Frank, A. (eds.) Geographic objects with indeterminate boundaries, pp. 109–121. Taylor Francis, London (1996)
9. Mathes, A.: Folksonomies - cooperative classification and communication through shared metadata, http://adammathes.com/academic/
   computer-mediated-communication/folksonomies.html
10. Janowicz, K., Raubal, M., Schwering, A., Kuhn, W.: Special issue on semantic similarity measurement and geospatial applications. Transactions in GIS 12(6) (2008)
11. Kuhn, W.: Semantic engineering. In: Navratil, G. (ed.) Research Trends in Geographic Information Science. Lecture Notes in Geoinformation and Cartography, vol. 12, pp. 63–76. Springer, Berlin (2009)
12. Nilsson, N.J.: The physical symbol system hypothesis: Status and prospects. In: 50 Years of Artificial Intelligence, pp. 9–17 (2006)
13. van Glasersfeld, E.: On the concept of interpretation. Poetics 12(2/3), 207–218 (1983)

14. Genesereth, M.R., Nilsson, N.J.: Logical foundations of artifical intelligence. Kaufmann, Los Altos (1987)
15. Guarino, N.: Formal ontology and information systems. In: Guarino, N. (ed.) Formal ontology in information systems: Proceedings of the first international conference (FOIS 1998), Trento, Italy, June 6 - 8. Frontiers in artificial intelligence and applications, vol. 46, IOS Press, Amsterdam (1998)
16. Reddy, M.J.: The conduit metaphor: A case of frame conflict in our language about language. In: Ortony, A. (ed.) Metaphor and Thought, pp. 164–201. Cambridge University Press, Cambridge (1993)
17. Boyd, R.N.: Scientific realism (2002),
    http://plato.stanford.edu/entries/scientific-realism/
18. Putnam, H.: The meaning of 'meaning'. In: Putnam, H. (ed.) Mind, Language, and Reality. Philosophical Papers, vol. 2, pp. 215–271. Cambridge University Press, Cambridge (1979)
19. Smith, B.: Beyond concepts: Ontology as reality representation. In: Varzi, A., Vieu, L. (eds.) Proceedings of FOIS, pp. 73–84. IOS Press, Amsterdam (2004)
20. Boyd, R.N.: On the current status of the issue of scientific realism. Erkenntnis 19, 45–90 (1983)
21. Lakoff, G.: Women, fire and dangerous things: What categories reveal about the mind. Univ. of Chicago Press, Chicago (1990)
22. Ladyman, J.: Understanding Philosophy of Science. Routledge, London (2002)
23. Quine, W.V.O.: Pursuit of Truth. Harvard University Press, Cambridge (1990)
24. Duhem, P.: The Aim and Structure of Physical Theory. Princeton University Press, Princeton (1954)
25. Quine, W.V.O.: Two dogmas of empiricism. The Philosophical Review 60, 20–43 (1951)
26. Mitchell, T.M.: The need for biases in learning generalizations. Technical report, Rutgers Computer Science Department, Technical Report CBM-TR-117 (1980)
27. Sober, E.: What is the problem of simplicity? In: Zellner, A., Keuzenkamp, H., McAleer, M. (eds.) Simplicity, Inference, and Modelling, pp. 13–32. Cambridge University Press, Cambridge (2002)
28. Cohen, S.M.: Aristotle's metaphysics (2008),
    http://plato.stanford.edu/entries/aristotle-metaphysics/
29. Anonymous: Matter, http://en.wikipedia.org/wiki/Matter
30. Flores, F.: The equivalence of mass and energy (2007),
    http://plato.stanford.edu/entries/equivME/
31. Bridgman, P.W.: The Nature of Physical Theory. Princeton University Press, New York (1936)
32. Eco, U.: Kant and the Platypus. Essays on Language and Cognition. Vintage, London (2000)
33. Quine, W.V.O.: Empirical content. In: Quine, W.V.O. (ed.) Theories and Things, pp. 24–30. Harvad University Press, Cambridge (1981)
34. Quine, W.V.O.: Word and object, 24th pr. edn. MIT Press, Cambridge (2001)
35. Grice, H.: Logic and conversation. In: Cole, P., Morgan, J.L. (eds.) Speech acts. Syntax and Semantics, vol. 3, pp. 41–58. Academic Press, New York (1975)
36. Quine, W.V.O.: The Roots of Reference. Open Court Publishing, LaSalle, Ill (1973)
37. Putnam, H.: Reason, Truth and History. Cambridge University Press, Cambridge (1981)
38. Searle, J.R.: Social ontology: Some basic principles. Anthropological Theory 6, 12–29 (2006)

39. Jackendoff, R.: What is a concept, that a person grasp it? In: Languages of the Mind: Essays on Mental Representation, pp. 21–52. The MIT Press, Cambridge (1993)
40. Sloman, A.: Interactions between philosophy and artificial intelligence: The role of intuition and non-logical reasoning in intelligence. Artificial Intelligence 2, 209–225 (1971)
41. Palmer, S.E.: Fundamental aspects of cognitive representation. In: Rosch, E., Lloyd, B. (eds.) Cognition and Categorization, pp. 259–303. Lawrence Erlbaum Associates, Hillsdale NJ (1978)
42. Suppes, P., Zinnes, J.L.: Basic measurement theory. In: Luce, R.D. (ed.) Handbook of mathematical psychology, ch. 1-8, vol. 1, pp. 1–76. Wiley, New York (1967)
43. Kuhn, W.: Semantic reference systems. Int. J. Geogr. Inf. Science 17(5), 405–409 (2003)
44. Gibson, J.J.: The ecological approach to visual perception. Houghton Mifflin, Boston (1979)
45. Scheider, S., Janowicz, K., Kuhn, W.: Grounding geographic categories in the meaningful environment. In: Hornsby, K.S., Claramunt, C., Denis, M., Ligozat, G. (eds.) COSIT 2009. LNCS, vol. 5756, pp. 69–87. Springer, Berlin (2009)

Towards a Semantic Representation of Raster Spatial Data

Rolando Quintero, Miguel Torres, Marco Moreno, and Giovanni Guzmán

Intelligent Processing of Geospatial Data Lab-Centre for Computing Research-National Polytechnic Institute, Mexico City, Mexico
{quintero,mtorres,marcomoreno,jguzmanl}@cic.ipn.mx
http://geo.cic.ipn.mx
http://www.cic.ipn.mx

Abstract. In this work, a methodology to semantically describe spatial objects within a Raster Spatial Data Set is outlined. This approach attempts to describe the objects contained in the raster data. For example, in Digital Elevation Models (DEM), as case study, we propose to find out landforms contained in the model, giving a description like *"In this model there is a mountain having a maximum altitude of 302 meters and located between 19.09383° N and 99.85541° W; also there is a plateau having ...".* This methodology consists of three stages: conceptualization for describing the domain of knowledge to be represented; synthesis for extracting objects from spatial data; and description for representing the objects found in the knowledge domain. The work is focused on establishing the guidelines to semantically process raster spatial data, according to the properties, relationships and concepts involved in the context of the landforms for DEMs.

1 Introduction

In this work, a methodology to semantically describe semantically spatial objects within a Raster Spatial Data Set (RSDS) is outlined, particularly in Digital Elevation Models (DEM). We attempt to make a description based on the knowledge that people have about spatial things, like things that we can see on a landscape. The methodology consists of three stages: conceptualization, synthesis and description.

The conceptualization stage tries to capture the knowledge about the domain of problem. In other words, it is necessary to find and define concepts used while people talk or think about landforms. In practical terms, conceptualization has three parts: 1) conceptualization of the geospatial domain (high level), 2) conceptualization of the particular domain (landforms in this case) and 3) conceptualization of the application domain (by means of description to make). The GEONTO-MET methodology for making the conceptualizations of these levels is proposed.

Mainly, it is based on minimizing axiomatic relations, which allow us to move the remaining relationships to the conceptualization, giving to them more semantic richness. As part of case study, an ontology of geographic domain based on the data dictionaries of National Institute of Statistics, Geography and Informatics of Mexico (INEGI) that we called *Kaab* Ontology has been developed. Similarly, the dictionary

K. Janowicz, M. Raubal, and S. Levashkin (Eds.): GeoS 2009, LNCS 5892, pp. 63–82, 2009.
© Springer-Verlag Berlin Heidelberg 2009

of Spanish Royal Language Academy is used to define concepts of the landforms domain; as result, the ontology called Hunxeet is obtained.

The synthesis stage is the numeric one; many algorithms for extracting features from the RSDS were developed. The stage used commonly digital image processing approaches to treat DEMs, having phases of pre-processing, processing and post-processing. As result of this stage, parts of the RSDS, called "extracts" are obtained. Each extract is considered an instance of a concept.

The description stage determines what an "extract" is, and builds its semantic representation. The stage is carried out using the conceptualization that indicates which properties of an "extract" must be measured in order to consider it an instance of a certain concept. The use of templates is proposed to fulfill the schema according to the measurements.

The paper is organized as follows; Section 1 includes a brief description about the work realized. Also, previous works related to the state of the art are presented in this section. The proposed methodology and the stages that compose it are described in Section 2. The results of the stages are depicted in Section 3. Finally, conclusions and future works are pointed out in Section 4.

1.1 Previous Works

In this work, DEMs as a case study for the methodology proposed is used. In the state of the art, many works are guided from a numeric point of view and several numbers of them are focused on the flow analysis and extraction of drainage lines [1], [2], [3].

Also, other areas related to landform analysis and processing; particularly the geo-morphometry have been deeply studied, but they are always focused on a numeric approach [2], [4], [5]. However, some works used "categories" or "classes" for making landform analysis.

Other important works are focused on methods and methodologies for building on-tologies. The first method is described by Uschold and King [6], which was extended by Uschold and Grüninger [7]. Authors proposed some guidelines based on their experience in developing Enterprise Ontology. This ontology was developed as a part of the Enterprise Project. To build an ontology, according to Uschold and King's methodology.

Based on the TOVE (TOronto Virtual Enterprise) project, Grüninger and Fox [8], proposed a formal approach to building and evaluating ontologies. This methodology has been used to build the TOVE ontologies. It is inspired by the development of knowledge-based systems using first order logic. They proposed intuitively identify-ing the main scenarios, that is, possible applications in which the ontology will be used. Thus, a set of natural language questions, called *competency questions* is used to determine the scope of the ontology. These questions and their answers are both used to extract the main concepts and their properties, relationships and formal axioms of the ontology. It transforms informal scenarios in computable models.

The KACTUS approach described in [9] to investigating the feasibility of knowl-edge reuse in complex technical systems and the role of ontologies to support it is proposed. It is conditioned by application development. Thus, every time an applica-tion is built, the ontology that represents the knowledge required for the application is refined. The ontology can be developed by reusing others and can be integrated in ontologies of later applications.

The METHONTOLOGY approach proposed [10], [11] the enabling of ontologies construction at the knowledge level. It has its basis in the main activities identified by the software development process and in knowledge engineering methodologies. This methodology includes: the identification of the ontology development process, a life cycle based on evolving prototypes, and methods to carry out each activity in the management, development, and support tasks.

The SENSUS-based-method is proposed by [12] for building the skeleton of the domain ontology, starting from a huge ontology. The SENSUS links domain-specific terms to the huge ontology and prunes in the huge ontology, those terms that are irrelevant for the new ontology. The result of this process is the skeleton of the new ontology.

The aim of On-To-Knowledge methodology is focused on applying ontologies to electronically available information in order to improve the quality of knowledge management in large and extended organizations. This approach proposes to build the ontology taking into account how the ontology will be used in further applications. Consequently, ontologies developed with this methodology are highly dependent on the application. A prior characteristic, it proposes ontology learning for reducing the efforts made to develop the ontology [13].

In [14] the Ontoclean, as a method to analyze and clean the taxonomy of an existing ontology by means of a set of principles based on the Philosophy is pointed out. It is oriented to remove wrong *Subclass-Of* relationships in taxonomies, according to some philosophical notions such as *rigidity*, *identity* and *unity*. These are applicable to properties, but can be extended to concepts.

On the other hand, some works related to geo-ontologies construction and semantics in geospatial information science have been developed.

For example, in [15] is reported the results of a series of experiments designed to establish how non-expert subjects conceptualize geospatial phenomena. Subjects were asked to give examples of geographical categories in response to a series of differently phrased elicitations. The results yielded an ontology of geographical categories – a catalogue of the prime geospatial concepts and categories shared in common by human subjects, independently of their exposure to scientific geography.

In [16] is designed an ontology of geographic kinds to yield a better understanding of the structure of the geographic world, and to support the development of GIS that is conceptually sound. This work demonstrated that geographical objects and kinds are not only larger versions of the everyday objects and kinds previously studied in cognitive science.

Methodologies and approaches described in this section are used with several purposes: to create a new ontology from scratch, to enrich an existing one with new terms, and to acquire knowledge for some tasks. Ontologies aim to capture *consensual* knowledge of a given domain in generic and formal ways to be reused and shared across applications and by groups of people.

2   The Proposed Methodology

The concept of spatial semantics (SS) has been treated from a general point of view to make semantic description of any type of spatial data. Moreover, our approach for SS definition is pointed out as follows: *"The semantics of a set of spatial objects is given by definition and/or description of those objects according to a conceptualization of the domain, in which objects have been processed"* [17], [18], [19].

It seems to be evident the necessity to specify a conceptualization for each particular case study. For instance, in [20] is outlined the application of SS for processing vector data; it is made giving an ontology of geographic domain and an application ontology for describing and structuring a spatial database. In [19], the SS is used for controlling quality in the map automatic generalization process.

According to these works, a semantic representation of spatial data that is composed of three stages: conceptualization, synthesis and description, which are used to process semantically raster spatial data sets.

2.1   Conceptualization Stage

The GEONTO-MET approach [17], [18], [20] is oriented towards formalizing geographic domain conceptualization according to specifications of the INEGI.

The main goal is to provide semantic and ontological descriptions, which represent the properties and relationships being described so that the behavior and features of geographic objects are taken into account directly from the geographic domain ontology.

The GEONTO-MET is composed of four principal tasks: *Analysis* provides an abstract model of the geographic objects involved in this domain. *Synthesis* carries out the conceptualization of the geographic domain. A set of application ontologies (in tourist and topographic contexts) and domain ontology called *Kaab-Ontology* is generated by the *Processing* stage. Finally, *Description* produces an alternative representation of geographic objects as well as the integration of them in a semantic description template.

Basically, the approach is based on a set of axiomatic relations allowing for direct translation of the relationships between concepts within the conceptualization. With this mechanism, the *semantic resolution* is improved, because the definition of such relationships can be iteratively refined. In other words, it offers a higher *semantic granularity* to the conceptualization and it is more flexible for generating ontological descriptions. For this, a couple of sets ($A_1 = \{is, has, does\}$ and A_2 specific prepositions related to geospatial context) are used. These sets are enough to define the rest of relationships, involved in geographic domain conceptualization, considering the definition of our methodology.

The essence of the approach is to reduce the *axiomatic relations* within the conceptualization. One could think that this reduction is a limitation for the richness of expression that conceptualization can implicitly contain; nevertheless, the universe of possible relations is not *a priori* defined, due to the fact that *"relation"* in a classic sense is not predefined. In fact, the reduction of axiomatic relations has two main advantages; the first, of which being the possibility of defining as many *"typical relations"* as needed, because this type of relationship is treated as a *concept*. In other words, *"typical relations"* are part of the conceptualization, there are not considered as axioms, they are defined as *concepts*.

The second advantage is that relations have a semantic association to themselves, not only from an axiomatic definition, but also from the conceptualization itself (the context of each relation).

To illustrate the concept, let us consider one widely used axiomatic relation: *"part_of"*. Such a relationship means that one concept is a constituent element of another concept.

With GEONTO-MET, it is possible to create this relationship as a concept *(concept-R)*, by defining the concept *"part"* (in the way that the concepts are defined) and using the axiomatic relations (*is, has, does*) to create the *concept-R* equivalent to the relationship *"part_of"*. For instance, let us consider the following sentence: **"heart** *part_of* **body"**; in this, two concepts (*heart* and *body*) are involved as well as one axiomatic relation (*part_of*).

By using the approach, the same relationship could be expressed **"heart** *is* part of **body"**, in which three concepts (*heart*, *part* and *body*) and two axiomatic relations: a fundamental one (*is*) and an auxiliary one (*of*) are described. The advantage is that the semantic of the relationship is the same and it is not necessary to previously define the relation *"part_of"*. Then, we only need to define *"part"* as a concept (having the semantic richness of concepts) and use it to define the new relationship.

On the other hand, GEONTO-MET is composed of a set of elements that are used to make *geographic domain conceptualization* that are described as follows.

Axioms Definition

This approach minimizes the number of axioms by means of the reduction of axiomatic relations. To make this process, a small set of axiomatic binary relations, divided into two subsets has been defined. The first, subset A_1 contains three relations that will be called *fundamental relations*. Equation 1 denotes the relations.

$$A_1 = \{is, has, does\} \tag{1}$$

The *"is"* relation means an *existence* or *identity*, such it is used to characterize the concepts in the conceptualization. It implies the inheritance of properties and abilities. Also, it allows for making a hierarchy of the same concepts. Some names for this relation could be *"son-of"* or, *"is-a"*. As a binary relation, *"is"* has the following properties.

Let C be a set of concepts:

- **It is anti-symmetric:** $\forall a,b \in C, a(is)b \wedge b(is)a \Rightarrow a=b$, if concept A *"is"* B and B *"is"* A, necessarily A and B are the same concept. For instance, "lake" *"is"* "waterbody" but "waterbody" *("is")* not lake, because there are other geographic concepts that are also "waterbody".
- **It is reflexive:** $\forall a \in C, a(is)a$, each concept A *"is"* itself. For instance, "Chapultepec Lake" *"is"* "Chapultepec Lake", because a geographic object has its own identity.

- **It is transitive:** $\forall a,b,c \in C, a(is)b \wedge b(is)c \Rightarrow a(is)c$, if A *"is"* B and B *"is"* C then A *"is"* C. For instance, if "road" *"is"* "og_artificial" and "og_artificial" *"is"* "geographic object", then "road" *"is"* "geographic object".

The *"has"* relation describes *aggregation* or *association*, such it is used to define the properties that build-up a concept. The properties of this binary relation are as follows:

- **It is non-symmetric:** $\exists a,b \in C, a(has)b \wedge b(has)a \Rightarrow a = b$, there are concepts A and B such that if A *"has"* B and B *"has"* A, then A and B are the same concept. For instance, "state" "has" "counties" but "counties" do not have (*"has"*) "states".
- **It is irreflexive:** $\exists a \in C \ni a\neg(has)a$, there is some concept A such that it does not have itself as property. For instance "urban area" does not have (*"has"*) "urban area".
- **It is transitive:** $\forall a,b,c \in C, a(has)b \wedge b(has)c \Rightarrow a(has)c$, if A *"has"* B and B *"has"* C, then A *"has"* C. For instance, if "country" *"has"* "state" and "state" *"has"* "county", then "country" *"has"* "county".

The *"does"* relation is used to describe an action, such it defines the *abilities* or operations associated with a concept. Its properties are as follows:

- **It is symmetric:** $\exists a,b \in C, a(does)b \wedge b(does)a \Rightarrow a = b$, there are concepts A and B such that if A *"does"* B and B *"does"* A, then A and B are the same concept.
- **It is irreflexive:** $\exists a \in C \ni a\neg(does)a$, there is some concept A that does not have itself as an ability.
- **It is non-transitive:** $\exists a,b,c \in C \ni a(does)b \wedge b(does)c \Rightarrow \neg a(does)c$, if A *"does"* B and B *"does"* C, it does not necessarily imply that A *"does"* C.

The second subset of axiomatic relations is denoted by A_2 and it is composed of prepositions. See Equation 2.

These relations are defined as *asymmetric*, *irreflexive* and *non-transitive*, although linguistically some of them do not accomplish such properties, for implementation convenience, we have considered that relations in A_2 are defined as follows.

$$A_2 = \begin{cases} to, before, under, with, against, of, from, in, between, towards, \\ until, for, by, since, on, after, behind, beside, near, through \end{cases} \quad (2)$$

Let $r \in A_2$ be a relation, this implies that:

i) $\forall a,b \in C, arb \wedge bra \Rightarrow a = b$
ii) $\exists a \in C \ni a\neg ra$ and,
iii) $\exists a,b,c \in C \ni arb \wedge brc \Rightarrow a\neg rc$.

Relationships Definition
Relationships in GEONTO-MET can be classified in two types:

- **Simple:** This type has the form: $a\rho b \in R_S$, where $a,b \in C^K$ and $\rho \in A_1$, C^K is the set of all concepts defined in the conceptualization K and R_C^K is the set of simple relationships for the conceptualization K.
- **Complex:** It has the form: $a\rho b\pi c \in R_c^K$, where $a,b,c \in C^K$, $\rho \in A_1$ and $\pi \in A_2$, and R_C^K is the set of complex relationships for the conceptualization K.

The *"is"* relation is a hierarchical relationship that provides the mechanism of *"inheritance"*. By using this relation, a hierarchy of concepts (existence) can be created. The *"has"* relation provides the capability of *"aggregation"* or *"composition"* of concepts, and the *"does"* relation describes the actions of a concept.

Relations in A_2 allow for describing and providing *causality* or *intention* to the relations in A_1. The complex form of relationships is used to have a minimum number of axiomatic relations and give us the semantic to non-axiomatic relationships. R_V describes the set of valid relationships. Also, it is the set of all relationships such that comply with restrictions. On the other hand, R_R represents the set of relationships existing in the real or concrete conceptualization.

Concepts Definition
In this approach a *concept* is defined as a collection of abilities and properties that share a single existence. There are four types of concepts:

- **Relational concepts** (verbs). They are defined as elements denoting an action or operation over other concepts. C_R represents the set of concepts-R.
- **Standard concepts** (noun). They are defined as elements belonging to a class. All their abilities and properties are abstract. C_E denotes the set of concepts-E.
- **Class concepts.** They are concepts-E that allow for making partitions of C_R and C_E. They are described by C_L (concepts-L).
- **Instance concepts.** They are concepts whose abilities and properties are concrete.

Also, we can say that $C = C_R \cup C_E \cup C_L$, such that $C_E \cap C_R = \emptyset$, $C_E \cap C_L = \emptyset$ and $C_L \cap C_R = \emptyset$, i.e. C_L, C_R and C_E are disjoined sets. An *abstract concept* is a concept-E, which does not have *instances*.

Properties Definition
Properties are concepts aggregated to one another by means of a relation of belonging. This defines the characteristics of the second concept. A property can be defined as follows:

Let $a,b \in C$ be concepts, we say that b is a property of a if $a(has)b \in R_R$. b is a concrete property of a if b is an **instance**. It is called *abstract property* when the concept is not concrete. For instance, if "mountain" *"has"* "altitude", and "altitude" could be "low", "mid" and "high"; then "altitude" is a property of "mountain" and "low" is a concrete property of "mountain".

Π is the set of properties, Π_C is the set of concrete properties and Π_A is the set of abstract properties. So, we can say that $\Pi = \Pi_A \cup \Pi_C$ and $\Pi = \Pi_A \cap \Pi_C = \varnothing$.

Abilities Definition

Abilities are concepts that define *actions* or *operations* associated with other concepts. Thus, they describe how a concept interacts with other concepts. An *ability* is defined as follows:

Let $a, b \in C$ be concepts, so we can say that b is an ability of a if $a(does)b \in R_R$.

b is a concrete ability of a if b is an **instance**. The non-concrete abilities are called *abstract abilities*. H is the set of abilities, H_C is the set of concrete abilities and H_A is the set of abstract abilities. Additionally, $H = H_A \cup H_C$ and $H_A \cap H_C = \varnothing$.

Instances Definition

Instances are a collection of concrete abilities and properties that have a unique existence. In other words, an *instance* is a concept contained within a hierarchy formed by the *"is"* relation, whose properties and abilities are concrete.

Thus, an *instance* is a concept with no sons within the hierarchy, - it is not a class and all its abilities and properties are instances. I represents the set of all instances in a conceptualization.

Abstract Classes Definition

Classes are concepts that have *children concepts* by means of the existence relation (*"is"*), their sons cannot be an *instance*, and they create a *complete partition* in the hierarchy.

These classes represent and classify the geographic objects from an abstract view and in a cognitive sense according to the domain experts. The *abstract classes* involve geographic objects considering their intrinsic characteristics. This partition can only inherit a set of subclasses that are defined as concepts-E. Subclasses are instanced to create concepts that describe each geographic object by means of axiomatic relations.

The *"is"* relation is equivalent to the relation *"subclass_of"*. In addition, C_L represents the set of classes. Table 1 defines the classes in the ontology and Table 2 sketches out the essential concepts (subclasses) that belong to abstract classes in the ontology.

Constraints Definition

They are statements explicitly defined to avoid inconsistencies in elements defined in the conceptualization. These constraints restrict the set R_V^K (set of valid relationships for conceptualization K). O_G denotes the ontology that conceptualize the geographic domain. O_D is the ontology that describes the specific domain.

These two ontologies are linked each other (and form a bigger one), so relationships must exist between concepts in O_D and concepts in O_G. For setting such relationships, the genealogy of a concept $a \in C^K$ as the set of concepts having an existence relation is defined. It can be expressed as follows. Let $a, b \in C^K$, then genealogy $G(a)$ is recursively given by $G(a) = \{b | a(\text{is})b \in R_R^K\} \cup G(b)$, where R_R^K is the set of real or concrete relationships for conceptualization K.

Also, the genealogy of a set of concepts is defined by $X = \{x_1, x_2, ..., x_n\}$, as the union of genealogies of each concept, it is denoted by Equation (3):

$$G(X) = G(\{x_1, x_2, ..., x_n\}) = \bigcup_{j=1}^{n} G(x_j).$$

Therefore, if $a \in O_D$, $b \in G(a)$ and $c \in O_G$, then $\forall a \exists c, b(is)c \in R_R^K$. In other words, all concepts in O_D are related to a concept in O_G directly or indirectly (through of its ascendants).

2.2 Synthesis Stage

The synthesis stage is used to decompose RSDS in "extracts" according to the conceptualization. In order to obtain "extracts", a strategy similar to the one used by people for clustering objects: applying different criteria to different clustering level is used.

For instance, if we had a set of furniture and tried to cluster different pieces, first we would separate chairs into a set, then the tables and finally the blackboards. In the next level of clustering different criteria for each group of furniture would be used.

For example, we would fix chairs by their color (brown, black), tables by their use (classroom, laboratory) and blackboards by their state (bad, good).

We will apply different clustering criteria to each set of objects. In each level, different algorithms or criteria are used. As result of the conceptualization of criteria and algorithms, the application ontology O_A is obtained.

As part of conceptualizing methodology an application model is stated. It is based on the basic statement: "algorithm produces extracts". By using GEONTO-MET, the statement is expressed as follows: "algorithm (*does*) produce (*to*) extract".

The conceptualization of the application produces the ontology with all possible results of applying the extraction algorithms. Moreover, concepts in O_D, as part of application conceptualization are linked establishing the existence relationships between extracts (unnamed concepts) in O_A and concepts in ontology O_D. It is denoted as follows.

Let $a \in O_A$ and $b \in G(a)$ then $\forall a \in O_A \exists c \in (O_G \cup O_D), b(\text{is})c \in R_R^K$.

2.3 Description Stage

The description stage consists of representing the "extracts" that have been obtained by the synthesis stage, according to the conceptualization. The "extracts" are described in the ontology O_A linked by the ontologies O_D and O_G.

However, it is not always possible to link an "extract" to the best concept in conceptualization; for example, to assign an "extract" to the concept "mountain" or the concept "hill", it is necessary to know how to differ between those concepts.

Let us review the definitions of such concepts: *"Mountain is a great natural elevation of land"*; *"Hill is a natural elevation of land lower than a mountain"*. Then, in this stage the properties of concepts to improve the linking of them to concepts in ontologies O_D and O_G must be analyzed.

In the example, we can stand that a mountain has higher elevation than a hill; so we must obtain the value of property "elevation" from the extract and use this to identify is the "extract" is a mountain or it is a hill. To determine the best specialization of an "extract", the clustering approach is used.

For instance, in Fig. 1a an "extract" belonging to O_A and linked to a concept in O_D that is associated with some properties (prop1, prop2 and prop3) is shown.

The concept has three children concepts (A, B and C). It must be obtained measuring the properties given by the conceptualization (1), obviously those measurements are taken from "extract" (2). As result, the properties of "extract" with their values assigned are obtained (3), (see Fig. 1b).

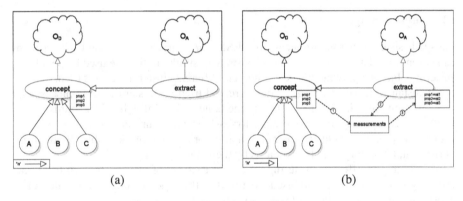

(a)   (b)

Fig. 1. Methodology for describing extracts

Fig. 2a illustrates the next stage of process, in which reference values for properties are obtained (4), this process will allow knowing to which of children concepts must belong the "extract".

Those reference values are obtained from the conceptualization (*a priori* knowledge) or by means of a training process (5). With these reference values, the "extract" is classified (6) and it is assigned to the best concept (7).

In the next step (Fig. 2b), the relation "is" is assigned (8) between "extract" and the best concept (B, in the example). Now, it is necessary to measure additional properties that the new concept could have (9).

In previous steps, measurements are taken from the "extract" (10). Finally, it fulfills its properties with the previous ones (11).

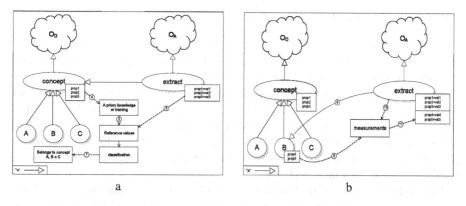

Fig. 2. Methodology for describing extracts (Cont.)

To make the description, the first "extract" identified as instance of a concept must be searched (step 1 in Fig. 3). This is made visiting the nodes of the "extract" in the hierarchical tree. Once that an "extract" is found, the description is started from (2).

At this point, the relationship that an "extract" has with a concept in O_D for obtaining existential information about the "extract" should be followed (2.1). It obtains the label of the concept as well as its properties and abilities (2.2 and 2.3).

Once the description of "extract" is made, we must search for next "extract" to be described (3) and repeat the description process (2) until no more "extracts" could be found in O_A (4).

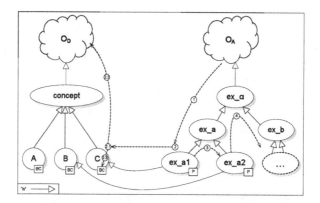

Fig. 3. Methodology for describing extracts (Cont.)

3   Experimental Results

As results, we used the methodology to semantically describe spatial objects within a Raster Spatial Data Set. Particularly, this methodology is used to describe Digital Elevation Models as case study.

3.1 Results of Conceptualization Stage

The conceptualization has been carried out in three parts: conceptualization of geographic domain, landforms domain and application. By applying the methodology three ontologies (*Kaab*, *Hunxet* and *Wiinkil*) were designed.

Kaab Ontology. In Fig. 4, a fragment of the *Kaab* ontology with the main classes defined to conceptualize the geographic domain is depicted. Detailed description of each class as well as all remaining concepts are presented in [17] and [18]. For this conceptualization, the topographic 1:50000 vector data dictionary for Mexico is used [21], in which more than 70 topographic features are defined and translated to the conceptualization presented in [17] and [18].

Hunxeet Ontology. In Fig. 5 some concepts of landforms in ontology that we have called *Hunxeet* are outlined. This conceptualization is based on the dictionary of the Spanish Royal Academy of Language.

Wiinkil Ontology. The application ontology, that we call *Wiinkil*, is a conceptualization in form of hierarchy of "extracts" obtained from the extraction algorithms. In the case of landforms, an algorithm that gives three types of "extracts": depressions, elevations and plains is used. In Fig. 6 depicts this hierarchy.

Finally, the developed ontologies are integrated to use and enrich the knowledge described by them.

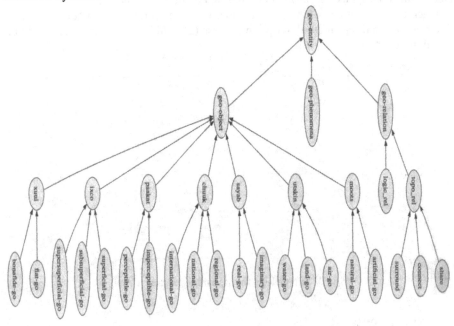

Fig. 4. A fragment of Kaab Ontology

Fig. 5. A fragment of Hunxeet Ontology

Fig. 7 shows the integration of Kaab ontology with Hunxeet ontology by means of the assignation of the main class in *Hunxeet* (landform) to corresponding classes in *Kaab*: "perceptible-go", "real-go", "land-go", "superficial-go", "regional-go", "natural-go" and "bonafide-go". In this way the landforms are characterized, according to the ontology of the geographic domain.

3.2   Results of Synthesis Stage

The synthesis algorithm generates three types of extracts: "elev" for elevations, "depr" for depressions and "llan" for plains. By applying the algorithm recursively, different combinations of these types are generated. The specific combination is called *signature*.

The algorithm consists of four steps: 1) compute the longer plain zone (ZLE), 2) region labeling, 3) segmentation and 4) extraction. For computing ZLE, the second derivate of DEM is used and the Laplace filter is applied.

With the second derivate the longest region containing only zeros (by using a region growing algorithm with 8-conex neighborhood). Once ZLE is found, the next step is to separate data according to their relative altitude respect to ZLE.

Elevation data below minimum altitude of ZLE are labeled as "depr", data above maximum altitude of ZLE are labeled as "elev"; the remaining data is labeled as "llan".

The segmentation step is done separating data labeled and obtaining three "new" RSDSs: one with "elev" signature, other with "llan" and the third with "depr" as its signature.

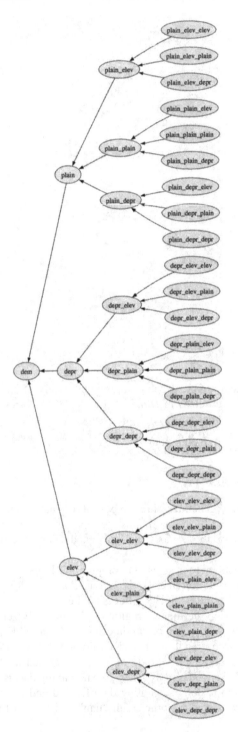

Fig. 6. A fragment of Wiinkil Ontology

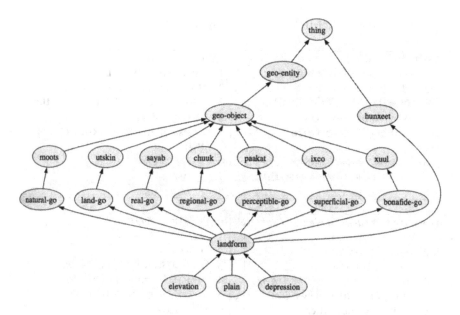

Fig. 7. Kaab-Hunxeet integrated ontology

Finally, in order to obtain the resulting "extracts", the areas (8-connected) within RSDS are separated. Table 1 describes the pseudo-code for the synthesis stage.

For testing the algorithm, a DEM obtained from USGS web site of Grand Canyon on Colorado State, USA is used. Fig. 8 shows the result of segmentation step. Here, four little images: labeled with "zle" are shown (the Largest Plain Zone).

They are labeled with "elev", "llan" and "depr" obtained from data sets classified under corresponding signature. According to the conceptualization, we must obtain 21 data sets (21 signatures), before starting the extraction step. Fig. 9 depicts the results of the extraction step. In the figure, each extract identified by different color is visualized.

3.3 Results of Description Stage

In the description stage, the "extracts" obtained are semantically represented. In Table 2 is presented the algorithm for describing an "extract", where *ext* y *sign* are the "extract" and its signature is presented.

Also, GET-PROPS is a procedure to determine the properties of the "extract" (1), according to the conceptualization and the signature. The procedure MEASURE-PROPS computes the values of the properties directly from elevation data contained in the "extract" (2). GET-TEMP is a procedure that obtains the template that corresponds to the "extract" and the signature (3).

Finally, the template is fulfilled with existence information, as well as with property values by means of the procedure FILL-TEMPL (4). In Table 3 the result of the description stage is presented.

Table 1. Pseudo-code for Synthesis Algorithm

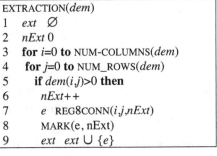

SEARCH-ZLE(dem)
1   lap LAPLACE-FILTER(dem)
2   zle ∅
3   **for** i=0 **to** NUM-COLUMNS(lap)
4   **for** j=0 **to** NUM_ROWS(lap)
5   **if** lap_{ij} = ZERO() **then**

6         $zone$ REG8CONN(lap,i,j)
7   **if** IS-BIGGER($zone,zle$) **then**
8   zle $zone$

LABELING(zle,dem,lbl)
1   **for** i=0 **to** NUM-COLUMNS(dem)
2   **for** j=0 **to** NUM-ROWS(dem)
3   **if** dem_{ij}< MIN-ALTITUDE(zle) **then**
4   etq_{ij} "depr"
5   **else if** dem_{ij}> MAX-ALTITUDE(zle) **then**
6   etq_{ij} "elev"
7   **else**
8   etq_{ij} "llan"

SEGMENTATION($dem,lbl,elev,llan,depr$)
1   $elev$ NO-DATA()
2   $llan$ NO-DATA()
3   $depr$ NO-DATA()
4   **for** i=0 **to** NUM-COLUMNS(dem)
5   **for** j=0 **to** NUM_ROWS(dem)
6   **if** lbl_{ij}="elev" **then**
7         $elev_{ij}$ dem_{ij}
8   **else if** lbl_{ij}="llan" **then**
9         $llan_{ij}$ dem_{ij}
10   **else if** lbl_{ij}="depr" **then**
11        $depr_{ij}$ dem_{ij}

EXTRACTION(dem)
1   ext ∅
2   $nExt$ 0
3   **for** i=0 **to** NUM-COLUMNS(dem)
4   **for** j=0 **to** NUM_ROWS(dem)
5   **if** $dem(i,j)$>0 **then**
6         $nExt$++
7   e REG8CONN($i,j,nExt$)
8   MARK($e, nExt$)
9   ext ext ∪ {e}

Table 2. Pseudo-code for Description Algorithm

DESCRIPTION($ext,sign$)
1   $props$=GET-PROPS($sign$)
2   $vals$=MEASURE-PROPS($ext,props$)
3   $plant$=GET-TEMP($sign$)
4   $desc$=FILL-TEMP($plant,vals$)
5   **return** $desc$

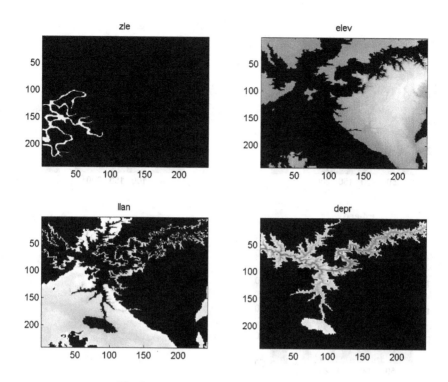

Fig. 8. First iteration of the segmentation step

Table 3. Results of Description Stage

```
RASTER SPATIAL DATA SET OF DIGITAL ELEVATION MODEL FROM "GRAND CANYON - E
AZ       ", HAVING SPATIAL RESOLUTION OF 30 SECONDS-ARC. MAX ALTITUDE:
548.000000 METERS, MIN ALTITUDE: 2838.000000 METERS. EXTREME COORDS:
(129600.000000, -406800.000000) AND (133200.000000, -403200.000000) SECOND-
ARC, PROJECTION: GEOGRAPHIC. THIS RSDS HAS:
A MOUNTAIN WITH AREA: 168300 SQUARE SECONDS-ARC, MIN ALTITUDE: 2699.301185
METERS, MAX ALTITUDE: 2774.861966 METERS, EXTREME COORDS: (132540.000000, -
405660.000000) AND (136680.000000, -401460.000000) SECOND-ARC, TOP:
(132930.000000, -405540.000000) AT A HEIGHT OF 2774.861966 METERS
...
A HILL WITH AREA: 7200 SQUARE SECOND-ARC, MIN ALTITUDE: 2640.503674 METERS,
MAX ALTITUDE 2650.418398 METERS, EXTREME COORDS: (133050.000000, -
405750.000000) AND (136800.000000, -402090.000000) SECOND-ARC, TOP:
(133170.000000, -405720.000000) AT A HEIGHT OF 2650.418398 METERS...
```

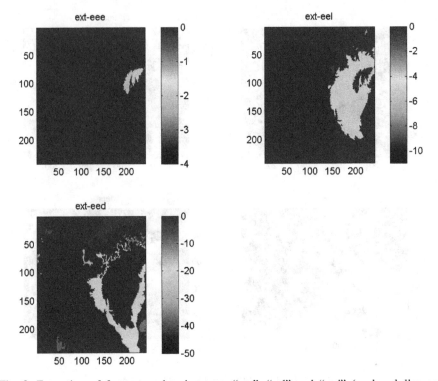

Fig. 9. Extraction of features under signatures "eee", "eel" and "eed" (e=elev, l=llan and d=depr)

4 Conclusions

In this work, a methodology for making semantic descriptions of raster spatial data sets is described. The conceptualization methodology is the most important part of this research; because we propose to make the conceptualization using only three axiomatic relations, which allow to move the "classic" relationships to the conceptualization, giving to them a granularity and semantic richness.

As part of case study three ontologies: *Kaab* ontology for the conceptualization of geographic domain, *Hunxeet* ontology for the conceptualization of landforms domain, and *Wiinkil* ontology for the conceptualization of our application were developed.

Synthesis stage is focused on the image processing fashion, with phases of pre-processing, processing and post-processing. *Description* stage is proposed to use the conceptualization and apply some templates for describing geospatial knowledge.

As future work, we consider that it is necessary to analyze and conceptualize geographic relationships (topologic and geometric for instance) between concepts identified and described in this work. Also, it is important to consider methods for measuring the quality of the description.

We propose the use of building blocks (basic landforms) for building a synthetic model and compare it to the original data set. On the other hand, the description by using formal first order logic and comparing the resulting logics, in order to obtain a quality metric will be proposed.

Acknowledgments

The authors of this paper wish to thank the CIC, SIP Projects: 20082563, 20082580, 20082480, 20080971, 20091264, 20090320, 20091018, 20090775, IPN and CONACYT for their support.

References

1. Ackermann, F.: Automatic generation of digital elevation models, OEEPE Commision B, DTM Accuracy Meeting, Southampton (1993)
2. Hodgson, M.E.: What cell size does the computed slope / aspect angle represent? Photogrammetric Engineering and Remote Sensing 61(5), 513–517 (1995)
3. Etzelmüller, B., Sulebak, J.R.: Developments in the use of digital elevation models in periglacial geomorphology and glaciology. Physische Geographie 41, 35–58 (2000)
4. Weibel, R., DeLotto, J.L.: Automated terrain classification for GIS modeling. In: Proceedings of GIS/LIS 1988, Virginia (1998)
5. Sulebak, J.R., Tallaksen, L.M.: Estimation of areal soil moisture by use of terrain data. Geografiska Annaler 82(A), 89–105 (2000)
6. Uschold, M., King, M.: Towards a Methodology for Building Ontologies. In: Proceedings of the Workshop on Basic Ontological Issues in Knowledge Sharing IJCAI 1995, Montreal, Canada, pp.6.1–6.10 (1995)
7. Uschold, M., Grüninger, M.: Ontologies: Principles, Methods and Applications. Knowledge Engineering Review 11(2), 93–155 (1996)
8. Grüninger, M., Fox, M.S.: Methodology for the design and evaluation of ontologies. In: Proceedings of the Workshop on Basic Ontological Issues in Knowledge Sharing IJCAI 1995, Montreal, Canada, pp. 7.3–7.13 (1995)
9. Bernaras, A., Laresgoiti, I., Corera, J.: Building and reusing ontologies for electrical network applications. In: Proceedings of European Conference on Artificial Intelligence, Budapest, Hungary, pp. 298–302. John Wiley & Sons, Chichester (1996)
10. Fernández, M., Gómez, A., Juristo, N.: METHONTOLOGY: From Ontological Art Towards Ontological Engineering. In: Symposium on Ontological Engineering of AAAI, pp. 33–40. Standford University, California (1997)
11. Gómez, A., Fernández, M., Corcho, O.: Ontological Engineering, 2nd edn. Springer, New York (2004)
12. Swartout, B., Ramesh, P., Knight, K., Russ, T.: Toward Distributed Use of Large-Scale Ontologies. In: Symposium on Ontological Engineering of AAAI, pp. 138–148. Standford University, California (1997)
13. Staab, S., Schnurr, H.P., Studer, R., Sure, Y.: Knowledge Processes and Ontologies. IEEE Intelligent Systems 16(1), 26–34 (2001)
14. Guarino, N., Welty, C.: A Formal Ontology of Properties. In: Dieng, R., Corby, O. (eds.) EKAW 2000. LNCS (LNAI), vol. 1937, pp. 97–112. Springer, Heidelberg (2000)
15. Smith, B., Mark, D.: Ontology and Geographic Kinds. In: Proceedings of the 8th International Symposium on Spatial Data Handling, Vancouver, Canada, pp. 308–320 (1998)

16. Mark, D.M., Smith, B., Egenhofer, M., Hirtle, S.: Emerging Research Theme: Ontological Foundations for Geographic Information Science. University Consortium for Geographic Information Science, Technical Report (2001)

17. Quintero, R.: Representación Semántica de Datos Espaciales Raster, Laboratorio de Procesamiento Inteligente de la Información Geoespacial, México, CIC-IPN, Ph. D. Thesis in Spanish (2007)

18. Torres, M.: Representación ontológica basada en descriptores semánticos aplicada a objetos geográficos, Laboratorio de Procesamiento Inteligente de la Información Geoespacial, México, CIC-IPN, Ph. D. Thesis in Spanish (2007)

19. Moreno, M.: Similitud Semántica entre Sistemas de Objetos Geográficos Aplicada a la Generalización de Datos Geo-espaciales. Laboratorio de Procesamiento Inteligente de la Información Geoespacial, México, CIC-IPN, Ph. D. Thesis in Spanish (2007)

20. Torres, M.: Ontological representation based on semantic descriptors applied to geographic objects. Computación y Sistemas 12(3), 356–371 (2009)

21. INEGI: Diccionario de datos topográficos 1:50 000 (Vectorial), Aguascalientes, Instituto Nacional de Estadística Geografía e Informática (1996)

Bottom-Up Gazetteers: Learning from the Implicit Semantics of Geotags

Carsten Keßler, Patrick Maué, Jan Torben Heuer, and Thomas Bartoschek

Institute for Geoinformatics, University of Münster, Germany
{carsten.kessler,patrick.maue,jan.heuer,
bartoschek}@uni-muenster.de

Abstract. As directories of named places, gazetteers link the names to geographic footprints and place types. Most existing gazetteers are managed strictly top-down: entries can only be added or changed by the responsible toponymic authority. The covered vocabulary is therefore often limited to an administrative view on places, using only official place names. In this paper, we propose a bottom-up approach for gazetteer building based on geotagged photos harvested from the web. We discuss the building blocks of a geotag and how they relate to each other to formally define the notion of a geotag. Based on this formalization, we introduce an extraction process for gazetteer entries that captures the emergent semantics of collections of geotagged photos and provides a group-cognitive perspective on named places. Using an experimental setup based on clustering and filtering algorithms, we demonstrate how to identify place names and assign adequate geographic footprints. The results for three different place names (*Soho*, *Camino de Santiago* and *Kilimanjaro*), representing different geographic feature types, are evaluated and compared to the results obtained from traditional gazetteers. Finally, we sketch how our approach can be combined with other (for example, linguistic) approaches and discuss how such a bottom-up gazetteer can complement existing gazetteers.

1   Introduction and Motivation

The amount of geotagged user-generated content on the Social Web has been soaring in the last years. Cheaper and smaller GPS chips as well as easy-to-use tools for manual geotagging have led to a sharp increase, particularly in the number of geotagged photos. The sheer amount of geotagged pictures – currently over 100 million on Yahoo's Flickr service alone[1] – makes them a very attractive source for geographic information retrieval [1,2]. As such, geotagged photos can be regarded as an implicit kind of Volunteered Geographic Information (VGI) [3]. Merging professional data sources with such VGI is attractive for a number of reasons, such as rapid updates and enrichment with data typically not contained in professional data sets. Examples include the extraction of footprints [1] and grounding of vague geographic terms [4] such as *downtown Mexico City*

---

[1] According to http://blog.flickr.net/2009/02/05/

K. Janowicz, M. Raubal, and S. Levashkin (Eds.): GeoS 2009, LNCS 5892, pp. 83–102, 2009.
© Springer-Verlag Berlin Heidelberg 2009

or mapping of non-geographic terms [5] to determine the regional use of words like *soda* or *pop* [6].

One promising use of VGI – and geotagged photos in particular – is the enrichment of gazetteers with vernacular names and vague places [7]. Gazetteers have been developed as directories of named places with information on geographic footprints and place types to facilitate geographic information organization and retrieval. Most gazetteers follow a strict top-down approach, i.e., the gazetteer data is administered by the organization running the gazetteer. Only this toponymic authority can add places or place types to the gazetteer and correct erroneous entries, which slows down updates and hampers the inclusion of local and often tacit knowledge. Moreover, in most gazetteers information on geographic footprints is limited to a single coordinate pair, representing the centre of a city, administrative district or street. Extraction of footprints from geotagged information on the web is thus a promising way to automatically generate polygonal footprints for these gazetteer entries. Although a number of approaches have been developed for this task [5,8,9,10], they are hardly implemented in existing gazetteers. Apart from the GeoNames gazetteer[2], which complements its database with geotagged information from Wikipedia, strict top-down management of gazetteers is still prevalent.

In this paper, we present an approach to build gazetteers *entirely* from volunteered geographic information. We discuss the challenges posed by automatically establishing the foundations of such a gazetteer based on geotagged photos harvested from the web. The implemented algorithms for retrieving geotags and clustering the corresponding locations to generate footprints are well-established. However, the *emergent semantics* [11] of such a collection of geotagged photos is still largely unspecified. Hence, the main contribution of this paper will be the formal definition of geotags. We explain the relation between the attached label (tag) and the information objects like a photo, its label's author, as well as creation time and coordinates. We discuss the implicit semantics hidden in this relation, and how gazetteer entries can emerge from collections of such geotags using the presented implementation.

Inferred knowledge about places from a source like geotagged photos – usually tagged with subjective keywords – can be seen as a social knowledge building process [12, chapter 9]. Ideally, this process leads to a representation of the *group cognition* [12] and can thus be regarded as a cognitive engineering [13] process which lets traditional GI applications benefit from the *Wisdom of the Crowds* [14]. Gazetteers exposing the collaborative perspective on place differ significantly from traditional gazetteers with administrative focus [15]. It is thus not the aim of this research to replace today's gazetteers, which have already proven useful for countless applications building on geocoding, geoparsing and natural language processing. Instead, we argue for a separation of these different views into separate gazetteers, which can then be accessed through a gazetteer infrastructure as outlined in [7,16].

---

[2] See http://www.geonames.org

In order to demonstrate the feasibility of our approach, we have set up an application which retrieved and processed geotags associated to photos published on Flickr, Panoramio and Picasa[3]. While there is also other geotagged content online such as videos, blog posts or Wikipedia entries, we chose to limit this experiment to photos. Photos are inherently related to the real world, since every photo has been taken *somewhere*. Moreover, as mentioned above, there is already a substantial amount of geotagged photos available online. By analyzing the coordinate pairs attached to the pictures, the time they were taken as well as the tags added by their owners, we are able to compute geographic footprints representing specific keywords. The collection of these keywords, derived from all tags of all retrieved photos, is further analyzed to differentiate between toponyms and tags without spatial relation. We test a repository build up this way with queries for *Soho, Camino de Santiago* (Way of St. James) and *Kilimanjaro*. We compare the results to those obtained from the same query on GeoNames. This evaluation focuses on the question whether our bottom-up gazetteer can already take on established gazetteers in terms of completeness and accuracy of geographic footprint.

The next section points to relevant related work. Section 3 introduces a formal definition of geotags and establishes the relation between gazetteers and geotags. Section 4 describes the crawling and filtering approach implemented in the prototype. Section 5 analyzes the results obtained for the three exemplary queries, followed by conclusions and an outlook on potential applications and future work in Section 6.

2 Related Work

This section points to related work from gazetteer research, tagging and bottom-up generation of geographic information.

2.1 Gazetteer Building and Learning

Gazetteers are knowledge organization systems that consist of triples (N, F, T), where N corresponds to the place name, F to the geographic footprint and T to the place type [17]. Since neither N, F nor T are unique, all three components are required to fully represent and unambiguously identify a named place [17, p. 92]. In the context of gazetteers, a clear distinction is made between place as a social construct based on perceivable characteristics or convention [18], and the actual real-world feature it refers to [19]. Feature types are mostly organized in semi-formal thesauri with natural language descriptions. Recent research demonstrates how gazetteers could benefit from more rigorous, formal place type definitions [16] and develops methods for gazetteer conflation [20].

Existing gazetteers have generally been developed based on databases provided by administrative authorities, or by merging existing gazetteers [17]. More recently, the ever-growing amount of information available on the web has been

---

[3] See http://flickr.com/, http://panoramio.com/ and http://picasaweb.com/

identified as a promising resource of knowledge about named places. Jones et al. [1] introduce a linguistic approach to enrich gazetteers with knowledge about vague places. They use documents harvested via web search and analyze them for cooccurrences of vague place names with more precise co-located places. In another linguistics-based approach presented by Uryupina [21], a bootstrapping algorithm is applied to automatically classify places into predefined categories (e.g. *city*, *mountain*). The machine learning techniques employed in this research enabled a high precision of about 85%, albeit the comparably small training data sets of only 100 samples per category. Henrich and Lüdecke [5] introduce a process based on the results retrieved from a web search engine to derive geographic representations for both geographic and non-geographic terms at query time. Goldberg et al. [22] developed an agent-based system that crawls structured online source such as the USPS zip code database and online phone books. The authors demonstrate that this approach is capable of creating detailed regional, land-parcel level gazetteers with a high degree of completeness.

2.2   User-Generated Geographic Information

Online mapping tools with open APIs such as *Google Maps* have enabled the creation of the huge amounts of user-generated geographic information – also dubbed collaborative [23] or volunteered GI (VGI) [3] – in the first place. While this mainly refers to projects like OpenStreetMap[4], we argue that geotags, and more importantly the geographic footprints derived from them, can also be filed into this category. Similar approaches have already been sketched in previous research to derive landscape regions [24] or imprecise definitions of boundaries of urban neighborhoods [8] from such geotagged content. We build on this previous work and show how geographic information collected this way can be processed for the integration with existing gazetteers.

3   What Is a Geotag?

We have introduced geotags as particular examples of volunteered geographic information. Before discussing the idea of inferring semantics from the *geotag*, we are going to formally define it.

3.1   Tagging Geographic Information Objects

Humans adding items like pictures to their collections use individual ordering schemes (besides time) to group similar items, keep different items apart and consequently simplify recovery. We order books in our (real) book shelf according to various criteria, including topic, age, thickness, or even color. Such individual preferences re-appear in virtual collections. Using tags – words or combinations of words people associate with virtual items – is a well accepted approach to sort items on the virtual shelf. Tags, however, can vary significantly from person

---

[4] See http://www.openstreetmap.org/

to person. The formal definition of a tag therefore has to include both the user and the tagged information object. Gruber [25] suggests to model the tag as the process $Tagging = (L, U, I, S)$, which establishes an immediate relation between the the Label L coming from the User's (U) vocabulary associated to an information Item I. This definition includes a Source S, which enables sharing across applications. In the following, we leave this source aside, since it has no direct impact on the presented approach. The following rule states that, if a label is associated with an item by some user, it is regarded as tag. More importantly, it also states that a tag is always bound to its author and the item:

$$\forall l(Label(l) \land \exists i(Item(i) \land associatedTo(l,i))$$
$$\land \exists u(User(u) \land createdBy(l,u)) \rightarrow Tag(l) \tag{1}$$

Any information object which is inherently hard to classify – basically all non-textual information – requires a solution for its categorization. Tagging is commonly accepted for such contents, such as photos or videos, but also for bookmarks, scientific articles, and many more. In the remainder of this paper, we focus on photos with an identifiable geographical context, e.g. a picture of *La Catedral* in Mexico City. The items in question are therefore related to objects in the geographic landscape [26]. Goodchild's "geographic reality" [27] as formal definition of geographical information takes the spatio-temporal nature of the physical (field-based) reality into account. Humans, however, do not perceive reality as continuous fields. They identify individual objects, either directly or indirectly by looking at photos created by camera sensors.

In this *World of Individual Objects* [26] we only consider particulars (entities existing in space and time) with an observable spatial and temporal extension. Objects on the photo have per se no meaning; in Frank's *World of Socially Constructed Reality* we eventually associate semantics to be able to reference the particulars [28] in spoken language. Such reference can either be a proper name, which is used as unique identifier [29], e.g., *Catedral Metropolitana de la Ciudad de México*, or it links to a category[5] which groups objects sharing common properties, e.g. *cathedral*. We finally identify individual particulars according to their spatial or temporal characteristics, by either referring to complex objects (e.g., *downtown*) or to the homogenous spatial or temporal region the object is proper part of, e.g., *Mexico City*. So far, this follows the definition of gazetteer entries from Section 2.1. The place type T and place name N in the discussed triple (N, F, T) both refer to the particular's semantics, the geographic footprint F on the other hand is related to its spatial extension in physical reality.

The same applies to the labels used to tag a photo, which function as references to particulars in geographic space. The nature of this reference, however, cannot be explicitly described: although it appears to be obvious for the mentioned proper names or category names, most tags associated to photos do not have an objective relation to the geographic object. The label `vacation09` makes perfect sense for the user, who might have sorted all pictures of his Mexico trip using this

---

[5] The reference is then again the proper name of the object's type.

tag. Once the items are shared, however, such personal tags loose any usefulness. Other examples which have no immediate relation to the depicted particular are labels naming properties of the item itself (e.g. blue, high-resolution), the process of creating the item (e.g. nikon), its potential use (e.g. wallpaper), or simply the author's opinion (interesting). Note that we assume that it is the user's intention to improve the item's findability; hence, we do not expect to encounter deliberate errors (which is obviously not true in real world settings; we propose an effective solution for this problem in Section 3.3). Once we have identified the references, we can use them to locate the referred-to object in space and time. The following rule makes this dependency between the tag and its role as reference to the depicted particular explicit:

$$\forall l \exists i (\mathit{Tag}(l) \land \mathit{Item}(i) \land \mathit{associatedTo}(l, i) \land \tag{2}$$
$$\exists p (\mathit{Particular}(p) \land \mathit{represents}(i, p)) \rightarrow \mathit{refersTo}(l, p))$$

The rule does not (and cannot) further specify the reference type. Taking our example of the cathedral, the label Catedral Metropolitana is immediately referencing – here as proper name – the particular. We can then further specify the tag as a proper name:

$$\forall l \exists p (\mathit{Tag}(l) \land \mathit{Particular}(p) \land \mathit{names}(l, p) \rightarrow \mathit{ProperName}(l)) \tag{3}$$

The open question here is obviously how to infer if the label is a proper name and, even more important, how to ensure that it is really the proper name of the depicted geographic object. The clustering and filtering approach introduced in the next sections provides answers to both questions.

Labels like Mexico or Summer 2009 are indirect references. They point to a region containing the particular (spatially and temporally, respectively). The following rule formalizes our assumption, that, if the tag is a toponym referring to a certain geographic region, we can infer that our depicted object is spatially related to that region:

$$\forall l \exists p (\mathit{Tag}(l) \land \mathit{Particular}(p) \land \mathit{refersTo}(l, p) \land \tag{4}$$
$$\exists r (\mathit{GeographicRegion}(r) \land \mathit{names}(l, r)) \rightarrow \mathit{spatiallyRelated}(p, r))$$

We can only assume that there is a spatial relation between the depicted particular and the place name. By looking only at the labels we cannot infer what kind of spatial (or temporal, for that matter) relation exists, and hence what spatial character this specific label has. In the following section we introduce the concept of a geotag as an extension of the traditional tag. Geotags give us the opportunity to make use of geographic coordinates and points in time to identify the spatio-temporal character of the associated labels.

3.2   A Formal Definition of Geotag

The tagging process establishes the relation between the user, the information item, and the label. If the information item represents one or more geographic objects, the associated label may (but does not have to) refer to either dimension of the depicted object: either its semantics (including a proper name of the individual or category) or its spatio-temporal extension (naming, for example, the containing region). A geotag extends the notion of the tag by adding an explicit location in space and time to the information item. In the case of digital photos, a time stamp with the creation date is usually added by the camera automatically. Geographic coordinates are either provided by built-in GPS modules, or added manually by the user. Building on Gruber's definition of tagging as a relation, we add the time stamp T and the coordinates C to the relation (and omit the source S): $Geotagging = (L, U, C, I, T)$. By extending our rule-based definition of a tag (Eq. 1), the following rule reclassifies a label as a geotag

$$\forall l \exists i (Label(l) \wedge Item(i) \wedge associatedTo(l, i) \tag{5}$$
$$\wedge \exists c (Coordinate(c) \wedge associatedTo(c, i))$$
$$\wedge \exists t (Timestamp(t) \wedge associatedTo(t, i))$$
$$\wedge \exists u (User(u) \wedge createdBy(l, u)) \rightarrow Geotag(l))$$

Note that we do not assume that a label reclassified as geotag is per se a place name. The tag blue is not necessarily related to the depicted object, nor does it have a spatial or temporal character. In our understanding, it is still a geotag, since it is the label used by one user in some occasion to tag an item with an associated location and date. In the following Section 3.3, we introduce an approach which reliably computes whether a label is spatially related to the particular.

3.3   A Clustering Approach to Categorize Geotags

The definition of geotags introduced in the previous section has substantial implications on the conceptual level. An information item is linked to a coordinate and time stamp, and labelled by one or more individuals. If we want to extract one particular aspect, e.g. the spatial coverage of geotags, we have to consider the other four properties as well.

Using the definition of a geotag as the relation $Geotagging = (L, U, C, I, T)$, we use the *tuple relational calculus*[6] [30] in the remainder to specify the queries used to retrieve different kinds of *clouds*. For example, the query $\{g.C | g \in Geotagging \wedge g[L] = L_i\}$ returns the coordinates of all tuples g where the label (the field L) has the value L_i. We call the result of this query a point cloud of a label. A folksonomy – the aggregation of all tags from all users into one (uncontrolled) vocabulary – is then simply formalized as $\{g.L | g \in Geotagging\}$. The resulting tag cloud can also be reduced to the vocabulary of one particular user U_i with the query $\{g.L | g \in geotags \wedge g[U] = U_i\}$. Her spatio-temporal activity – the user's movement across space and time – is queried using the statement $\{g.C, g.T | g \in Geotagging \wedge g[u] = U_i\}$.

---

[6] TCR is a concise declarative query language for the relational model, the presented examples can also be expressed in SQL.

We suggest to make use of the point cloud of one label to compute its spatial footprint. A gazetteer build on top of this approach could then return geometries and centroids for proper (potentially unofficial) names of geographical objects. The information we derive from geotags, however, is inherently noisy: many tags do not have an immediate relation to the particular represented by the geotagged item. Only *significant occurrences* of geotags should therefore be considered for this approach. We define one occurrence of a geotag $g = (L_i, U_i, C_i, I_i, T_i)$ as significant if the following two conditions are fulfilled:

1. At least two tuples g_i and g_j exist where $g_i[L] = g_j[L]$, and $g[U_i] \neq g[U_j]$. Since names in geotags are subjective, this rule assures that only names which are used by different persons are taken into account.
2. The spatial distribution $\{g.C | g \in Geotagging \wedge g[L] = L_i\}$ can be clustered.

In the following section we describe the algorithm which applies filters checking for these conditions to extract the relevant candidates for toponyms from the large set of tags. The semantic analysis of the two preceding sections can be easily realized as executable rules, for example expressed in the Semantic Web Rule Language (SWRL) [31]. SWRL supports built-ins, the algorithm presented in the following pages can therefore be integrated as *geotag:significant* and used to extend and clarify rule 2:

$$\forall l \exists i (Tag(l) \wedge Item(i) \wedge \tag{6}$$
$$associatedTo(l, i) \wedge geotag : significant(l) \wedge$$
$$\exists p(Particular(p) \wedge represents(i, p)) \rightarrow refersTo(l, p))$$

A reasoning engine triggers the execution of the clustering algorithm once it processes the added built-in. The algorithm returns true if the given label is significantly occurring (or false otherwise). Once we have applied the filtering and clustering, our gazetteer can provide the point clouds (and the regions covered by the point clouds) for given place names. For some place names, the clustering process results in multiple clusters (see the example of *Soho* in the following sections). This does not impair the efficacy of the presented approach as long as the clustering algorithm produces reasonable results (which depends mostly on the number of available geotags). For cases such as *Soho*, multiple gazetteer entries are generated.

Although we introduced time as a fundamental component of the geotag, we have not discussed the implications for the targeted gazetteer. With the presented approach, the tag GEOS 2007 would also be classified as place name. While we cannot discuss this issue here in detail for a lack of space, distinguishing between toponyms and labels naming temporal events can be implemented by applying the clustering approach both to the spatial and temporal dimensions.

3.4   Extraction of Gazetteer Entries

Section 2.1 defines gazetteer entries as triples (N, F, T). This notion has to be further specified for a gazetteer based on geotags. Since, in our case, the underlying data consist of a large collection of photos geo-located with exactly one coordinate pair, the given place name N maps to a point cloud as geographic footprint: $F = \{g.C | g \in geotags \wedge g[L] = L_i\}$. Each point in the cloud represents one significant occurrence of the given place name as tag for a photo. Since the footprint is no longer a single coordinate pair, the gazetteer's mapping from place name to footprint $N \longrightarrow F$ should now result in three different mappings. $N \longrightarrow F_r$ maps the place name to the *raw* footprint consisting of the corresponding point cloud. $N \longrightarrow F_p$ maps to the *polygon* which approximates the region occupied by the point cloud. $N \longrightarrow F_c$ finally maps a place name to the footprint's *centroid*, i.e., to a single coordinate pair as returned by conventional gazetteers. The centroid is the mean of *all* coordinate pairs in the point cloud and is thus specifically (and intentionally) biased towards areas that contain high numbers of geotags. F_c can thus be regarded as the point of interest best representing a place name, based on the number and location of corresponding geotags.

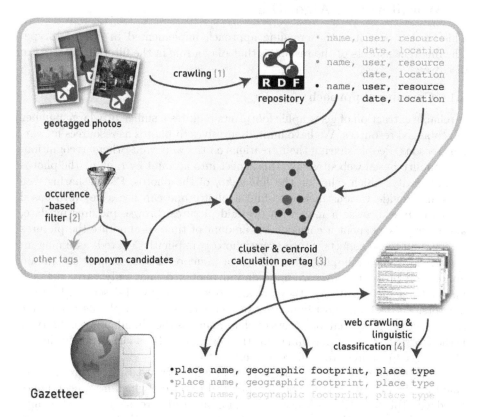

Fig. 1. Geotagged photos are crawled from the web (1) and fed into an RDF triple store. The tags are filtered based on occurrences to retrieve a subset of toponyms (2). For each place name, regions and centroids are calculated (3). Finally, every place name is categorized using linguistic classification (4). The part outlined in grey has been implemented for this paper (adapted from [7]).

While the derivation of the gazetteer entries from geotags allows for enhanced functionality in the mapping from place name to footprint, the mapping to place type $N \longrightarrow T$ remains unchanged. The experimental setup presented in Section 4 leaves the place type unspecified. Potential combinations with linguistic approaches [21] as sketched in Figure 1, however, would allow for a semi-automatic classification of the gazetteer entries based on a predefined typing scheme. This scheme could be adopted from existing gazetteers. Due to the limited reliability of any data coming from such collaborative platforms, such an approach would at least require quality control mechanisms. A fully automatic *strong* typing of place names with such bottom-up approach is clearly not feasible here. While this is out of scope for this paper, the grouping of a resource's tags into place names, place types and other tags does appear feasible. Moreover, it stands to reason whether such a tag-based typing is a more practical approach for a community-driven gazetteer [32].

4   Workflow and Algorithm

This section describes the crawling approach implemented in our prototype. The different aspects of the resources that play a role in the filtering process are discussed.

4.1   Crawling Approach

A reliable extraction of geographic footprints requires a sufficiently large number of geotagged resources. We have limited ourselves to photos as resources for various reasons. People sharing their creations on the web want others' recognition. Community-based web sites take this aspect into account by ranking the photos by popularity, which relies on the *findability* of the photos. Photo-sharing web sites all provide various means to find a photo: one can use a keyword-based search engine, browse a map with overlaid pictures, browse pictures by date, and so on. Users spent a considerable amount of time to annotate the pictures to cover all these aspects. Since every photo is implicitly located, assigning an explicit location by linking the photo to a point on a base map is a common annotation procedure. Accordingly, digital photos do not only carry detailed metadata in their Exif tags, they are also exceptionally well described by their creators. The last and most important reason to consider only photos as resource for extracting the spatial footprints of place names is the abundant availability. It is therefore reasonable to assume that the crawling yields a large enough sample of geotagged resources to achieve a significant result.

The crawling algorithm is conceptually straight-forward. Starting from a specific tag, the algorithm requests all geotagged resources which have been annotated with this tag. All three services used for our study provide this functionality through their APIs. For every tag attached to a retrieved photo, we store a separate complete geotag tuple (L, U, C, I, T) in our RDF triple store. In the next step, the conditions detailed in section 3.1 are applied to filter out tags which we have identified as not important. The resulting set of geotag tuples is taken as input for the clustering method described in the following.

4.2 Geotag Extraction Algorithm

A place name either refers to one unique place (e.g. *Kilimajaro*) or to multiple regions (e.g. the districts *Soho* in London and New York). The geotag tuples resulting from the crawling algorithm are used to identify clusters of high point-density. We consider the point cloud (explained in Section 3.3) as geographic footprint for the label L_i if many people used this keyword to annotate their photos taken nearby. Such clusters can have any shape, they are not necessarily concave and can contain holes. Point clouds derived from geotags are not equally distributed over space, but have some tendency to follow structures like trails or streets. In [10] the Delaunay triangulation has been identified as candidate algorithm to find clusters within point clouds. This method is not restricted to places with certain geometries. It computes the smallest possible triangle between three adjacent points; each point is connected to its nearest neighbors by an edge. A Delaunay triangulation for the tag *Soho* in New York is depicted in Figure 2. In order to split the graph of points and edges into clusters of high density (short edges), we remove all edges longer than a given threshold. If adjacent, remaining triangles are merged into one or more polygons. They represent F_p, the polygonal geographic footprint of the gazetteer's place name N.

A more advanced way to extract polygonal footprints from single locations is the Alpha Shape [33,34], which has also been used to generate the Flickr shape files[7]. For reasons of simplicity, we sticked to a Delaunay triangulation for this experiment. The next section shows that even with such a comparably simple clustering approach one can already obtain usable results.

Fig. 2. Cluster graph after the Delaunay triangulation for the place name *Soho*. The screen shot shows the clustering result depending on the edge length threshold: A small value results in several small clusters shown in blue. When the threshold increases, the fragments starts to join to the large black cluster.

---

[7] See http://code.flickr.com/blog/2008/10/30/

5 Experimental Results and Evaluation

This section presents the results obtained by our prototype implementation. The results are discussed and compared to those obtained from conventional gazetteers.

5.1 Soho, Camino de Santiago and Kilimanjaro

We retrieved geotagged photos annotated with *Soho, Camino de Santiago* and *Kilimanjaro*. These three place names were chosen because they represent different geometries: Soho as a city district represents polygonal real-world features up to a few kilometers in diameter. Moreover, we chose this example because there is not "the one" Soho, but both districts in London and New York can be regarded as equally well-known. Camino de Santiago refers to a number of pilgrimage routes leading to the Cathedral of Santiago de Compostela[8] in northwestern Spain. It usually refers to *Camino Francés*, the medieval route along Jaca, Pamplona, Estella, Burgos and León, but it is also used for a number of other ways to Santiago de Compostela across Europe and is thus a prime example of an ambiguous linear real-world feature. The third example, Kilimanjaro, is an example of a large-scale natural feature that can be seen (and hence shot) from far, but is hard to reach. Using this example, we want to investigate how well our approach is apt to derive useful results for such features.

Table 1. Figures on the RDF repository used for this study. The numbers include a negligible number of entries added during the testing phase.

Geotag Tupels	Filtered Geotag Tupels	Unique Names	Filtered Unique Names	Resources	Users
560,834	471,393	9,917	2,035	10,603	1,103

Table 1 gives an overview of the number of resources and tags obtained by crawling the three photo sharing websites for the three given examples. Only around 15 percent of tuples were removed during the filtering process, the ratio of ∼0.84 is surprisingly high. The ratio from filtered to unfiltered unique names on the other hand is ∼0.21; this shows that our filtering approach identified almost 80% of the names as irrelevant since they were used by only one user. The difference between the two ratios means that the remaining 20% of filtered uniques names appear in 80% of all geotag tuples. Our rather simple approach of not further considering tags that only occur once thus proves very effective. Most tags are noise, but those which remain are used and accepted by many users. Table 2 contains the specific numbers per place name.

For Soho, the two biggest clusters emerge as expected in central London and in New York (see Figure 3). Apart from these two main clusters, a number of

---

[8] Tradition has it that the cathedral contains apostle Saint James the Great's gravesite.

Table 2. Figures on the three case studies. The last column indicates the distance from the cluster's centroid to the corresponding footprint in GeoNames (a: London, b: New York).

Place name	Geotag Tuples	Resources	Users	Dates	Distance
Soho	11916	3124	446	3087	0.26^a / 0.16^b km
Camino de Santiago	5132	1304	75	1255	285.3 km
Kilimanjaro	2536	825	72	808	3.7 km

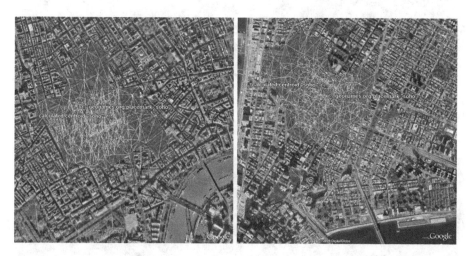

Fig. 3. The clusters generated for *Soho*. The left screen shot shows the cluster in London, the right one shows the cluster in Manhattan, New York.

smaller clusters appear at different locations around the world. An analysis of the corresponding resources showed that most of them correspond to smaller places called Soho, thus representing valid gazetteer entries. The small outlying clusters south of the main cluster in Figure 3, however, are clearly no meaningful results. Such outliers occur frequently when users tag whole photo sets with the name of the place where *most* of them were taken. This inevitably tags some photos with the wrong place name and will require an improved filtering approach.

For Camino de Santiago, the generated clusters give a good impression of the main trail to the Cathedral of Santiago de Compostela (see Figure 4). One apparent problem here is that the clustering algorithm splits up the route into distinct segments. Future research should focus on the development of "intelligent" clustering approaches that take the shape of the cluster into account, in order to enable a more reliable clustering.

For Kilimanjaro, the emerging clusters (see Figure 5) expose the main problem with an approach based on tagged and geolocated photos: users often do not tag the picture with the place name of the location where the picture was taken, but with the name of real-world feature *shown* in the picture. This becomes

Fig. 4. The clusters generated for *Camino de Santiago* give a good impression of the trail of the route

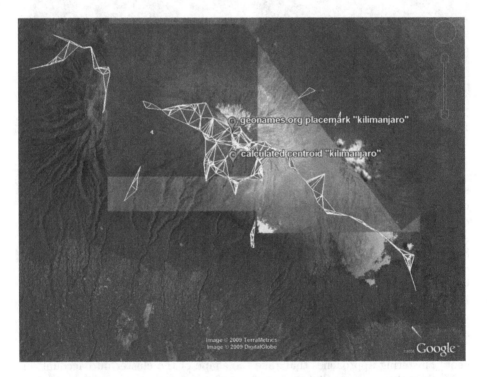

Fig. 5. The clusters generated for *Kilimanjaro* are distributed over a large area and show the problem of photos tagged with with the names of features shown in the pictures, although they were taken from far away

especially apparent for very large real-world features, as in this example. Several smaller clusters expose the high number of pictures taken at these locations, which apparently offer a good view on Mount Kibo, the highest peak of the Kilimanjaro massif. Future work needs to investigate how clusters referring to such real-world features can be detected, for example, by identifying ring-shaped clusters such as the one in Figure 5.

5.2   Geographic Footprints

The footprints extracted by our approach provide additional useful information to the point-based footprints provided by conventional gazetteers. For comparison with GeoNames, we also computed the corresponding centroid as the mean of all coordinates in every cluster (or cluster group, as for Kilimanjaro). This centroid points to what can be described as a named cluster's *group-cognitive centre*. In contrast to the geometric centre point, it gives an estimate of the common point of interest of users providing the photos retrieved in the crawling step. In the following, we discuss the extracted footprints and how the group-cognitive centre and the geometric centre point differ for our three examples.

For Soho and Kilimanjaro, the distance between the GeoNames footprint and the centroid of our cluster is comparably small, given the respective scale of the cluster (and the size of the corresponding real-world feature). The footprint for Soho, London, in GeoNames is about 260m away from the centroid of our cluster. The cluster itself represents the common notion of Soho very well[9], although it extends across Oxford Street in the north, which is usually taken as Soho's northern border. The same applies to the eastern extension of the cluster; the southern and western extension match the common notion of Soho very well. Similar observations can be made for Soho, New York: The area that is commonly referred to as Soho[10] is completely covered, but the cluster exceeds the actual area in all four directions. This exceeding problem can probably be addressed by adjusting the cutoff length during triangulation and fetching more input data. The centroid of the cluster is only 160m away from the footprint of the corresponding GeoNames entry. The clusters generated for Camino de Santiago stretch very well along the actual trail of the route, despite the gaps discussed above. The calculation of the centroid shows that it is in most cases meaningless to represent linear real-world features by points. While the centroid represents a mean value for all coordinates in the clusters, the footprint from GeoNames is located at one end of the route. Selecting the destination of the pilgrimage trail as footprint certainly makes sense in this case (the coordinate refers to Santiago de Compostela), however, this selection will be completely arbitrary for linear features that lack such a clear destination (such as most roads). For Kilimanjaro, the clusters represent the areas with a *view* on the Kilimanjaro's highest peak, rather than the mountain itself (due to the problems discussed above). This also causes a distance of almost 4 km of the clusters' centroid to the GeoNames

---

[9] See http://en.wikipedia.org/wiki/Soho#Streets for comparison.
[10] See http://en.wikipedia.org/wiki/SoHo#Geography

footprint, which is nevertheless still within an acceptable range given the size of the real-world feature.

6   Conclusions

This section summarizes the paper and points to different applications of the approach presented in this paper, as well as directions for future work.

6.1   Discussion

In this paper, we have presented an experiment to test the feasibility of the idea to build a gazetteer completely from geotagged photos crawled from the web. We have introduced the theoretical foundations to capture the emergent semantics of geographic information extracted from geotagged resources on the web. A theoretically sound definition of a *geotag* has been introduced and related to the classical definition of a gazetteer. Using the implementation which clustered and filtered geotags of photos, we have demonstrated how the geographic footprint for a given place name can be derived.

The results of our queries for *Soho*, *Camino de Santiago* and *Kilimanjaro* showed that it is possible to derive meaningful geographic footprints from geotagged content, even with comparably simple clustering approaches. Both the footprints as well as their centroids shed a different light on named places than conventional gazetteers. As pointed out in [22], every gazetteer extracted from online information can only be as good as the information it builds on. However, our experiment has demonstrated that useful results can already be obtained with very straight-forward means to extract a group-cognitive perspective [12] on place names. Hence, we do not propose to replace existing gazetteers by our approach, but to complement them within a gazetteer infrastructure [7,16]. Further improvements can be expected from implementing models of trust in the harvesting process, which would allow for an estimation of the quality of the geotags used for clustering [7,23].

From a visual inspection, the generated regions were judged to be plausible representations of the place names' geographic footprints. Particularly, the algorithm showed the capability to recognize different places carrying the same name, as shown in the *Soho* example. Moreover, the filtering algorithm has successfully sorted the crawled tags into toponyms and other tags based on the notion of significant occurrences. The example of *Kilimanjaro* has shown that very large real-world features are problematic for our approach, since they often appear in the context of photos that show them, but that were taken far away from the actual feature. Evidently, the results could be improved by more sophisticated crawling, filtering and clustering approaches.

6.2   Applications

While the crawling approach presented in this paper has been developed with the recursive generation of a bottom-up gazetteer in mind, the underlying algorithms are also potentially useful in a number of other applications. The user component, for example, could be used to derive communities and their vocabulary by analyzing how groups of users tag certain real-world features. The temporal component has only been used to identify occurrences and to filter events that might corrupt the place name recognition. Instead of treating these filtered events as noise, however, one could also imagine an application that specifically looks for such events based on temporal clusters. This would enable an automatic calculation of geographic footprints for such events, which could eventually be merged into event gazetteers [35,36].

The fact that every resource carries a time stamp and a user's name can also be used to extract individual space-time prisms [37,38]. This may provide insight into real-world social interactions between the users of photo sharing platforms, such as "who travelled together" or "who went to this party". The implications for privacy, however, are obvious and would require a careful consideration of ethical issues. From this perspective, the photo sharing platforms used in this paper might require more fine-grained mechanisms to give their users control over what information they want to reveal to whom. One method to prevent automatic generation of such profiles would be to allow users to exclude specific metadata (or combinations of them) from access through the respective APIs.

6.3 Future Work

The next step in this research will be the combination of the filtering and clustering algorithm presented in this paper with linguistic web crawling approaches. This would facilitate to go beyond place names and their geographic footprints and also extract the corresponding place type, as demonstrated by Uryupina [21]. It is, however, unlikely that it will also be possible to extract a *strong* place typing from user tags. While straightforward types such as *city*, *street* or *river* may still be found frequently enough in the tags for a reliable extraction, it is unlikely that a user tags a picture taken in Soho with *section of populated place* – the associated feature class (i.e., place type) in GeoNames. However, same as for footprints and centroids, such a bottom-up typing scheme would reflect place types used in common language, as opposed to the often somewhat artificial administrative place types used in current gazetteers. This bottom-up approach should also allow for a more flexible categorization that does not force every named place into exactly one category [32] in order to fully capture the emergent semantics of collections of geotagged content. We also plan to extend the existing implementation to take the temporal nature of geotags into account. This eventually results in the identification not only of place names, but also of names of events and processes with a spatial character.

Acknowledgments

This research has been partly funded by the SimCat project (DFG Ra1062/2-1 and DFG Ja1709/2-2, see http://sim-dl.sourceforge.net) and the GDI-Grid project (BMBF 01IG07012, see http://www.gdi-grid.de). Figure 1 contains geotag icons under a Creative Commons license from http://geotagicons.com.

References

1. Jones, C.B., Purves, R.S., Clough, P.D., Joho, H.: Modelling vague places with knowledge from the web. International Journal of Geographical Information Science 22(10), 1045–1065 (2008)
2. Larson, R.R.: Geographic information retrieval and spatial browsing. GIS and Libraries: Patrons, Maps and Spatial Information, 81–124 (April 1996)
3. Goodchild, M.F.: Citizens as voluntary sensors: Spatial data infrastructure in the world of web 2.0. International Journal of Spatial Data Infrastructures Research 2, 24–32 (2007)
4. Bennett, B., Mallenby, D., Third, A.: An ontology for grounding vague geographic terms. In: Eschenbach, C., Gruninger, M. (eds.) Proceedings of the 5th International Conference on Formal Ontology in Information Systems (FOIS 2008). IOS Press, Amsterdam (2008)
5. Henrich, A., Lüdecke, V.: Determining geographic representations for arbitrary concepts at query time. In: LOCWEB 2008: Proceedings of the first international workshop on Location and the web, pp. 17–24. ACM, New York (2008)
6. McConchie, A.: The great pop vs. soda controversy (2002), http://popvssoda.com (last visited august 1st, 2009)
7. Keßler, C., Janowicz, K., Bishr, M.: An agenda for the next generation gazetteer: Geographic information contribution and retrieval. In: ACM GIS 2009, Seattle, WA, USA, November 4–6. ACM, New York (2009)
8. Wilske, F.: Approximation of neighborhood boundaries using collaborative tagging systems. In: Pebesma, E., Bishr, M., Bartoschek, T. (eds.) GI-Days 2008. ifgiPrints, vol. 32, pp. 179–187 (2008)
9. Guo, Q., Liu, Y., Wieczorek, J.: Georeferencing locality descriptions and computing associated uncertainty using a probabilistic approach. International Journal of Geographical Information Science 22(10), 1067–1090 (2008)
10. Heuer, J.T., Dupke, S.: Towards a spatial search engine using geotags. In: Probst, F., Keßler, C. (eds.) GI-Days 2007 – Young Researchers Conference. ifgiPrints, vol. 30, pp. 199–204 (2007)
11. Aberer, K., Mauroux, P.C., Ouksel, A.M., Catarci, T., Hacid, M.S., Illarramendi, A., Kashyap, V., Mecella, M., Mena, E., Neuhold, E.J., et al.: Emergent semantics principles and issues. In: Lee, Y., Li, J., Whang, K.-Y., Lee, D. (eds.) DASFAA 2004. LNCS, vol. 2973, pp. 25–38. Springer, Heidelberg (2004)
12. Stahl, G.: Group Cognition: Computer Support for Building Collaborative Knowledge (Acting with Technology). MIT Press, Cambridge (2006)
13. Raubal, M.: Cognitive engineering for geographic information science. Geography Compass 3(3), 1087–1104 (2009)
14. Surowiecki, J.: The Wisdom of Crowds. Anchor, New York (2005)
15. Schlieder, C.: Modeling collaborative semantics with a geographic recommender. In: Hainaut, J.-L., Rundensteiner, E.A., Kirchberg, M., Bertolotto, M., Brochhausen, M., Chen, Y.-P.P., Cherfi, S.S.-S., Doerr, M., Han, H., Hartmann, S., Parsons, J., Poels, G., Rolland, C., Trujillo, J., Yu, E., Zimányie, E. (eds.) ER Workshops 2007. LNCS, vol. 4802, pp. 338–347. Springer, Heidelberg (2007)

16. Janowicz, K., Keßler, C.: The role of ontology in improving gazetteer interaction. International Journal of Geographical Information Science 22(10), 1129–1157 (2008)
17. Hill, L.L.: Georeferencing: The Geographic Associations of Information (Digital Libraries and Electronic Publishing). MIT Press, Cambridge (2006)
18. Casati, R., Varzi, A.C.: Parts and Places. The Structures of Spatial Representation. MIT Press, Cambridge (1999)
19. Goodchild, M.F., Hill, L.L.: Introduction to digital gazetteer research. International Journal of Geographical Information Science 22(10), 1039–1044 (2008)
20. Hastings, J.T.: Automated conflation of digital gazetteer data. International Journal of Geographical Information Science 22, 1109–1127 (2008)
21. Uryupina, O.: Semi-supervised learning of geographical gazetteers from the internet. In: Proceedings of the HLT-NAACL 2003 workshop on Analysis of geographic references, Morristown, NJ, USA, Association for Computational Linguistics, pp. 18–25 (2003)
22. Goldberg, D.W., Wilson, J.P., Knoblock, C.A.: Extracting geographic features from the internet to automatically build detailed regional gazetteers. International Journal of Geographical Information Science 23(1), 93–128 (2009)
23. Bishr, M., Kuhn, W.: Geospatial information bottom-up: A matter of trust and semantics. In: Fabrikant, S., Wachowicz, M. (eds.) The European Information Society – Leading the Way with Geo-information (Proceedings of AGILE 2007), Aalborg, DK. Lecture Notes in Geoinformation and Cartography, pp. 365–387. Springer, Heidelberg (2007)
24. Guszlev, A., Lukács, L.: Folksonomy & landscape regions. In: Probst, F., Keßler, C. (eds.) GI-Days 2007 – Young Researchers Conference. ifgiPrints 30, pp. 193–197 (2007)
25. Gruber, T.: Ontology of folksonomy: A mash-up of apples and oranges. International Journal on Semantic Web & Information Systems 3 (2007), http://tomgruber.org/writing/ontology-of-folksonomy.htm (November 2005)
26. Frank, A.: Ontology for spatio-temporal databases. In: Sellis, T.K., Koubarakis, M., Frank, A., Grumbach, S., Güting, R.H., Jensen, C., Lorentzos, N.A., Manolopoulos, Y., Nardelli, E., Pernici, B., Theodoulidis, B., Tryfona, N., Schek, H.-J., Scholl, M.O. (eds.) Spatio-Temporal Databases. LNCS, vol. 2520, pp. 9–77. Springer, Heidelberg (2003)
27. Goodchild, M.F.: Geographical data modeling. Computational Geosciences 18(4), 401–408 (1992)
28. Saeed, J.I.: Semantics (Introducing Linguistics). Wiley-Blackwell (2003)
29. Searle, J.R.: Proper names. Mind 67(266), 166–173 (1958)
30. Codd, E.F.: A relational model of data for large shared data banks. Communications of the ACM 13(6), 377–387 (1970)
31. O'connor, M., Tu, S., Nyulas, C., Das, A., Musen, M.: Querying the semantic web with SWRL, pp. 155–159 (2007)
32. Shirky, C.: Ontology is overrated – categories, links, and tags. Essay (2005), http://shirky.com/writings/ontology_overrated.html
33. Edelsbrunner, H., Kirkpatrick, D., Seidel, R.: On the shape of a set of points in the plane. IEEE Transactions on Information Theory 29(4), 551–559 (1983)
34. Edelsbrunner, H., Mücke, E.: Three-dimensional alpha shapes. ACM Transactions on Graphics 13(1), 43–72 (1994)

35. Allen, R.: A query interface for an event gazetteer. In: Proceedings of the 2004 Joint ACM/IEEE Conference on Digital Libraries, pp. 72–73 (2004)

36. Mostern, R., Johnson, I.: From named place to naming event: creating gazetteers for history. International Journal of Geographical Information Science 22(10), 1091–1108 (2008)

37. Hägerstrand, T.: What about people in regional science? Papers in Regional Science 24(1), 6–21 (1970)

38. Miller, H.J.: A measurement theory for time geography. Geographical Analysis 37, 17–45 (2005)

Ontology-Based Integration of Sensor Web Services in Disaster Management

Grigori Babitski[1], Simon Bergweiler[2], Jörg Hoffmann[1], Daniel Schön[3],
Christoph Stasch[4], and Alexander C. Walkowski[4]

[1] SAP Research, Karlsruhe, Germany
{grigori.babitski,joe.hoffmann}@sap.com
[2] DFKI, Saarbrücken, Germany
Simon.Bergweiler@dfki.de
[3] Itelligence AG, Köln, Germany
daniel.schoen@itelligence.de
[4] Institute for Geoinformatics, Münster, Germany
{staschc,walkowski}@uni-muenster.de

Abstract. With the specifications defined through the Sensor Web Enablement initiative of the Open Geospatial Consortium, flexible integration of sensor data is becoming a reality. Challenges remain in the discovery of appropriate sensor information and in the real-time fusion of this information. This is important, in particular, in disaster management, where the flow of information is overwhelming and sensor data must be easily accessible for non-experts (fire brigade officers). We propose to support, in this context, sensor discovery and fusion by "semantically" annotating sensor services with terms from an ontology. In doing so, we employ several well-known techniques from the GIS and Semantic Web worlds, e.g., for semantic matchmaking and data presentation. The novel contribution of our work is a carefully arranged tool architecture, aimed at providing optimal integration support, while keeping the cost for creating the annotations at bay. We address technical details regarding the interaction and functionality of the components, and the design of the required ontology. Based on the architecture, after minimal off-line effort, on-line discovery and integration of sensor data is no more difficult than using standard GIS applications.

1 Introduction

Disasters may be caused by flooding, earthquakes, technical malfunctions, or terrorist attacks, to name a few. The efficient handling of such emergencies, i.e., the management of the measures taken to fight them, is a key aspect of public security. This is especially true in an increasingly tightly interlinked world, where problems in one area may quickly cause problems in connected areas. This phenomenon often causes disasters to exhibit an explosive growth, especially during their early stages. Defensive measures in such a stage are still premature, leading in combination with the explosive growth to what has been termed the "chaos-phase" [22]. Methods for shortening that phase are widely believed to be essential for limiting the damage caused by the disaster.

One of the characteristics of the chaos-phase is the overwhelming flow of information that must be managed by the defense organizations, such as fire brigades and

K. Janowicz, M. Raubal, and S. Levashkin (Eds.): GeoS 2009, LNCS 5892, pp. 103–121, 2009.

the police. Depending on the scale of the disaster, each organization establishes a *crisis team*, i.e., a committee of officers deciding which actions to take, and monitoring their execution. To come up with informed decisions, members of the crisis team must process an enormous amount of heterogenous information, such as messages from the public, feedback from own forces in the field or from partner organizations, and – last not least – Geospatial information such as weather conditions and water levels. Our focus herein is on the latter. Since not only is the amount of information huge, but also it must be evaluated in a situation of extreme stress and pressure, it is of paramount importance that the information can be accessed quickly and with complete ease.

In the SoKNOS project[1], we develop a service-oriented system facilitating amongst other things the integration of Geospatial information. This integration is realized in a Geographic Information component (GI Plugin), which offers functionalities to query data from several geospatial web services, to visualize the data in a map component, and to analyze the data through integrated GIS functionalities. Additional analyzing capabilities (e.g. simulations) can be intergrated by adding external processing services. The difficulty of integrating new information into the map depends on the form the information comes in. Our most basic assumption is that the information is encapsulated into Web services conforming with the standard specifications of the Open Geospatial Consortium (OGC). The integration of basic maps is realized through adding data from Web Mapping Services (WMS). Vector data (e.g. risk objects) can be accessed through Web Feature Services (WFS) and hence require the creation of suitable queries which poses serious challenges; indeed, given the stress and pressure of the targeted scenario, pre-specified queries are necessary.

An interesting and important middle ground are sensors, accessible through e.g. the Sensor Observation Services (SOS) as specified by the Sensor Web Enablement (SWE) initiative of the OGC. As sensor data is time-dependent, what the user needs to provide is, essentially, the desired Geographic area, the desired time interval, and the desired properties to be observed. The SOS specification lays the basis for doing so in an interoperable manner. Areas and time points are fully covered by standards. The main problems remaining are:

(I) For identifying observed properties, mediation is required between the terminology of the user and that of the Web service design.

(II) The user may not even know a technical term for the observed property she is looking for, necessitating an option to search by related terms.

(III) For fusing the information of several sensors, data transformation (e.g. units of measurement) is needed, and duplicate data needs to be detected and removed.

(IV) Sensors may become dysfunctional and in such case need to be replaced with suitable alternative sensors.

Characteristic properties of disaster management are that (II) and (IV) are likely to occur, that the number and types of required sensor informations are manifold, that the persons needing them act under high pressure, and that these persons have hardly any IT knowledge. Given this, (I)–(IV) constitute a serious difficulty.

---

[1] Service-oriented architectures supporting networks in the context of public security; http://www.soknos.de

In our work, we have developed and implemented a tool architecture that addresses (I)–(IV), up to a point where discovery and integration of sensor data is no more difficult than using standard GIS applications. The key technique is to make use of *semantic annotations* in a purpose-designed *ontology*. The technicalities will be summarized directly below, and detailed later on in the paper. First, we need to clarify that our approach encompasses a separate *service registration* activity, which contrasts with *service usage*. These correspond to the two fundamentally different phases in our domain, *off-line* (prior to the disaster) vs. *on-line* (during the disaster). On-line, pressured and hectic users need to comfortably discover and integrate sensor data. As the basis for that, our approach assumes that – off-line, in peace and with ample time – each service has previously been registered. Such registration means to acquire the service (finding it in the Web), to create a description including the semantic annotation, and to store that description within a local registry.[2] Apart from exploiting the off-line phase in a suitable preparatory way, the distinction between service registration and service usage also serves for decoupling these activities, allowing them to be performed by different people. The person performing the registration will also be associated with the fire brigade/police. But she may well have more IT knowledge than typical crisis team members. (That said, clearly, this person will not be a logics expert, so creating the semantic annotations needs to be reasonably easy; if it is not, then the effort for creating them is very likely to lead to non-acceptance anyhow.)

A commonly used definition is that *an ontology is a formal, explicit specification of a shared conceptualization* [7]. In our context, we define an ontology called *Geosensor Discovery Ontology* (GDO). The GDO defines a terminology suitable for describing sensor observations and related entities. Put in simple terms, the GDO contains:

(a) A taxonomy of phenomena, i.e., of properties that can be observed by sensors.
(b) A taxonomy of substances to which phenomena (a) may pertain.
(c) A taxonomy of Geographic objects to which phenomena (a) may pertain.
(d) The relations between (a), (b), and (c).

To ensure sustainable modeling, the GDO design follows the guiding principles of the DOLCE foundational ontology [16,5]. Simply put, DOLCE corresponds to a kind of widely accepted "best practice" for ontological modelling, serving to avoid common modelling flaws and shortcomings.

The semantic annotations associate, for a SOS service, each of the service's observed properties with a concept from (a). Clearly, these annotations are easy to create. Our architecture provides a simple user interface for doing so via drag-and-drop. In the obvious manner, the annotations solve problem (I). Since phenomena (a) are organized in a taxonomy (enabling us to find more general/more specialized sensors), the GDO also provides sophisticated support for problem (IV). Substances and Geographic objects are likely candidates a fire brigade officer will use as related terms, hence (b), (c), and (d) together serve to solve problem (II). Problem (III), finally, is solved by standard transformations and straightforward usage of the SOS output information.

It is also required to make the entire functionality easily accessible to the user. Our Graphical User Interface does so via standard paradigms, and intuitive extensions

---

[2] Hence the term "discovery" in this paper refers to finding a suitable sensor, on-line, in a (potentially huge) local registry, *not* in the Web.

thereof. For service discovery, the area of interest is marked by mouse movements as a rectangle on a map; the desired time points are given by manipulating the boundaries of a time interval; search in the GDO – which from the user's perspective corresponds to selecting the desired observations – is realized by text search combined with taxonomy browsing and following links (given by the relations between pairs of concepts in the ontology). Once services are discovered, fusing and displaying their data amounts to a single drag-and-drop action for the user. The architecture was successfully demonstrated to an evaluation team of German fire brigade and police officers, obtaining a very positive rating; we give some more details on this in Section 6.

The paper is organized as follows. Section 2 provides a brief background on the OGC Sensor Observation Service and the Semantic Web. Section 3 introduces concrete use cases that we will use for illustration. Section 4 covers our architecture, detailing after an overview the design of the GDO, the semantic annotations, as well as sensor discovery and fusion. Section 5 discusses related work, and Section 6 concludes with summary remarks and a discussion of open issues.

2   Background

We briefly give the most relevant background on the SOS service specification, and the Semantic Web domain.

2.1   Sensor Observation Service

The goal of the OGC Sensor Web Enablement initiative is to enable the creation of web-accessible sensor assets through common interfaces and encodings [2]. Therefore, the SWE initiative defines standards for the encoding of sensor data as well as standards for web service interfaces to access sensor data, task sensors or send and receive alerts. The Sensor Observation Service (SOS) is part of the SWE framework and offers a pull-based access to observations and sensor descriptions [18]. The SOS operations are grouped into three different profiles: the *core* profile for retrieving the service descriptions, sensor descriptions and observations; the *transactional* profile for registering new sensors and inserting new observations; the *enhanced* profile for offering additional service functionalities.

In this work, we focus on the basic operations of the SOS defined in the core profile. The core profile comprises the GetCapabilities, DescribeSensor and GetObservation operation. The GetCapabilities operation returns a service description of the service containing information about the supported operations and parameters as well as the observations which are provided, e.g. spatial and temporal extent of the observations, producing sensors and observed properties. Sensor metadata like sensor position, calibration information or sensor administrator can be retrieved using the DescribeSensor operation. The sensor descriptions are usually encoded in the Sensor Model Language (SensorML), a data model and XML encoding for sensor metadata [1]. The core operation of the SOS depicts the GetObservation operation. It offers the possibility to query observations filtered by spatial and temporal extent, producing sensors, certain observed properties, and/or value filters.

The Observations and Measurements (O&M) specification [3] is utilized by the SOS to encode the data gathered by sensors. It defines a model describing sensor observations as an act of observing a certain phenomenon. The basic observation model contains five components: The *procedure* provides a link to the sensor which generates the value for the observation.The *observedProperty* references the phenomenon which was observed. The *Feature Of Interest (FOI)* refers to the real world entity (e.g., a river) which was target of the observation. The time, when the observation was made, is indicated by the *samplingTime* attribute.The *result* element contains the observation value. The observation acts as a property value provider for a feature: It provides a value (e.g. 27 Celsius) for an observed property (e.g. temperature) of the FOI (e.g. weather station) at a certain timestamp. The location to which the observation belongs is indirectly referenced by the geometry of the FOI.

2.2 Semantic Web

In a nutshell (and as far as relevant for this paper), the Semantic Web community is concerned with the investigation of how *annotations* within a *formal language* can help with performing many tasks in a more flexible and effective way. Specifically, we are herein concerned with a form of *semantic service discovery*. The idea is that each Web service of interest is annotated with (an abstract representation of) its meaning – what does it do? – and services are discovered by matching this annotation against a discovery query – what kind of service is wanted? – given in the same logic. Since the annotations and queries, formulated relative to a formal domain model encoding complex dependencies, can be far more precise than free text descriptions, this approach has the potential to dramatically improve precision and recall.

Semantic discovery is, by the standards of the field, a long-standing topic in the Semantic Web. Earlier approaches were often based on annotating with, and reasoning about, complex logic languages such as 1st-order logic or rich subsets thereof. See e.g. [13] for a classical Desciption Logics formalization. Arguably, most of these approaches suffer from the prohibitive complexity of creating semantic annotations and discovery queries (and from the prohibitive computational complexity of the required reasoning). A more recent trend in the Semantic Web community is to use more "lightweight" approaches putting less of a burden on these activities, at the cost of reduced generality and power – the slogan being "a little semantics goes a long way" [8]. Our approach falls into this class, with carefully designed technology targeted at providing added value, while keeping the complexity at a level that will lead to actual acceptance by end users (fire brigades etc) in the relevant domain.

3 Example Scenario

In our example scenario, the floodwater level of the Rhine river in Germany rises immensely during a long lasting thunderstorm. Cologne and the industry park of Dormagen are affected by the flood. People have to be evacuated and organizations from other German federal states are called to support the disaster management. After a dike has broken and a chemical plant is flooded nearby the Rhine river, explosions occur which release pollutants into the air and the water. The emergency staff as well as residential areas around the chemical plant are threatened by the released air and water pollutants. We consider the following use cases for the proposed architecture:

(A) **Discovery and fusion of heterogenous water level measurements.** To get a more precise overview, all water gauges along the Rhine upstream of Cologne shall be integrated into the SoKNOS System. The sensor data is provided by different SOS services, using different identifiers for the observed phenomenon (e.g. water level, water gauge, gauge height), using different units of measurement, and partially overlapping each other. The challenges addressed by our architecture are to mediate between the identifiers and the terminology of the non-expert user, to make the sensors easy to find among a huge set of available sensors, to merge multiple data points, and to recognize redundant data.

(B) **Replacement of a water level measurement sensor.** The data displayed to the crisis team of course must be up-to-date. Since access to SOS services is pull-based, the map component sends new queries periodically. One of the sensors may have become damaged, and hence may now be out of order. The challenge addressed by our architecture is to recognize this, and to discover and integrate a suitable replacement sensor automatically.

(C) **Discovery and fusion of heterogenous air pollutant concentration measurements.** With conventional methods, the monitoring of air pollutant concentration is a time consuming and complicated task. There are only few vehicles with appropriate sensors. Hence the spatial resolution of the measured values is rather coarse grained. It takes considerable time for the vehicles to arrive at the area of interest, and the measurements are transferred through verbal communication, prone to delays and misunderstandings. This can be improved considerably through leveraging on resources – SOS services – that happen to be available in the particular scenario: the monitoring systems of chemical plants near the flooding. These SOS services could of course also be integrated off-line into conventional systems. But our approach allows to discover and use them with ease, based on minimal integration effort. Indeed, since registering a service requires hardly more effort than knowing where the service is and which phenomena it observes (see Section 4.3 below), it is conceivable that the integration is performed on-line, e.g. by a system administrator, upon demand by the crisis team members.

4   Semantic Sensor Integration

We now explain in detail our architecture, its individual components, and their design and functionality. We begin in Section 4.1 with an overview, giving a rough picture of the components and their interaction. We then delve into the details, describing in Section 4.2 the design of our ontology, explaining in Section 4.3 our semantic annotations and how they are created, describing in Section 4.4 our methods for sensor discovery, and describing in in Section 4.5 our methods for sensor data extraction and fusion. All user interactions are illustrated with screen shots, and all methods are exemplified with the use cases introduced in the previous section.

4.1 Architecture

Figure 1 shows an overview of our architecture. There are six components. Two of these are graphical user interfaces (GUIs, shown in the top left part of the figure), two are backend components (shown in the bottom left part), and two are data stores (shown on the right).

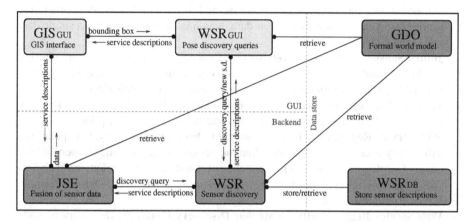

Fig. 1. An overview of our architecture

The Geographic Information System (GIS) GUI is basically a standard GIS map component, extended to cater for the required interactions with the Web Service Registry (WSR) GUI and the Joined Sensor Engine (JSE). The Web Service Registry GUI is the user interface of the Web Service Registry, which serves for registering and discovering Web service descriptions – i.e., descriptions of SOS services, including their semantic annotations, in our case. The Joint Sensor Engine extracts the data from a set of discovered services. It makes the required data transformations and it detects duplicate data. Most importantly, it monitors the performance of the services, and replaces them – via posing a suitable discovery query to the WSR – fully automatically in case of failure. The Geosensor Discovery Ontology (GDO) is a formalization of the domain, i.e., of the relevant terminology relating to sensor data, as outlined in the introduction. The Web Service Registry database (DB), finally, is the storage container for service descriptions. A brief summary of the interactions is as follows:

- **GIS GUI with Web Service Registry GUI.** The user specifies a bounding box via marking a rectangle on the map within the GIS GUI; the bounding box is sent to the Web Service Registry GUI, to form part of the discovery query. The discovery query is completed in the Web Service Registry GUI, and the discovered services are sent back to the GIS GUI. From that point on, the GIS GUI is responsible for displaying the data of these services.
- **Web Service Registry GUI with Web Service Registry.** Discovery queries are created in the Web Service Registry GUI, comprising the desired area (the bounding box), the desired time interval, as well as the desired kind of phenomenon to be

observed. The queries are sent to the Web Service Registry, which performs the discovery and sends the discovered service descriptions back to the Web Service Registry GUI. Additionally, the user may enter a new service description (possibly including a semantic annotation) in the Web Service Registry GUI, which is then sent to the Web Service Registry for storage.

- **GIS GUI with Joined Sensor Engine.** Whenever the GIS GUI needs to extract up-to-date data from the discovered sensors, it sends their descriptions to the Joined Sensor Engine. Based on the descriptions, the Joint Sensor Engine connects to the services, and extracts and fuses their data, which is then sent back (as a single data set) to the GIS GUI.
- **Joined Sensor Engine with Web Service Registry.** Whenever service monitoring inside the Joint Sensor Engine finds that a sensor has failed, it queries the Web Service Registry for replacement services, delivering equivalent data.
- **Web Service Registry with Web Service Registry DB.** The Web Service Registry connects to the database for storage and retrieval of service descriptions.
- **Web Service Registry GUI with Geosensor Discovery Ontology.** For specifying a discovery query, the user needs to find the desired concepts in the Geosensor Discovery Ontology, i.e., suitable phenomena or related entities. For that, the Web Service Registry GUI uses the structure of the Geosensor Discovery Ontology, which is read from the storage.
- **Web Service Registry with Geosensor Discovery Ontology.** Discovery is made not only directly on the concepts in the query, but also indirectly through the connections within the Geosensor Discovery Ontology, read from the storage.
- **Joined Sensor Engine with Geosensor Discovery Ontology.** For the purpose of data transformation, the Joined Sensor Engine needs information from the Geosensor Discovery Ontology in order to detect equivalent observed properties.

These functionalities and interactions will now be explained in detail. We start by detailing the structure of the GDO, which lies at the heart of our approach.

4.2   Ontology Design

The GDO is formalized in F-Logic [12], a logic based programming language which we chose mainly for practical reasons: F-Logic provides sufficient modelling power for our purposes, while at the same time being computationally efficient in the reasoning tasks we require.[3] In what follows, we do not delve into details of the formalization. Instead, we describe the design of the GDO at an intuitive level.

The GDO is designed to support discovery of SOS services, so, naturally, it builds on the relevant specifications [3,18]. SOS service descriptions contain keywords (called "observed properties" in (O&M) [3]) indicating the properties measured by the sensor. These properties are not standardized, but the CF Metadata[4] contains a (incomplete) collection. The GDO models those properties relevant for our application, as well as

---

[3] There is also a version of the GDO formulated in the standard description logic based language OWL [17]. In our work, this version mainly serves as a reference model. For the sake of simplicity, we do not discuss the OWL version and its relation to the F-Logic version.

[4] NetCDF Climate and Forecast (CF) Metadata Convention (http://cf-pcmdi.llnl.gov).

some supplementary entities, in the form of taxonomies of categories. Our technology connects those to real sensors via F-Logic rules.

An important aspect of the GDO is that it follows well-established ontological design principles. We align the GDO with the well-known DOLCE foundational ontology. DOLCE essentially is a kind of widely accepted "best practice" for ontological modelling. This serves to avoid common modelling flaws and shortcomings. For details regardng DOLCE, we refer the reader to the literature [16,5,6]. In what follows, a rough understanding of the following four concepts will suffice. **Endurants** and **perdurants** are distinct regarding their behavior in time. Endurants are wholly present at any time they exist, whereas perdurants extend in time by accumulating different temporal parts. Perdurants embrace entities generally classified as events, processes, and activities. An endurant "lives" in time by participating in some perdurant(s). For example, a building (endurant) participates in its lifespan (perdurant). In the GDO, we use two sub-categories of endurant: "non-agentive physical object" and "amount of matter". **Qualities** are the basic entities we can perceive or measure, for example the volume of a lake, the color of a rose, or the length of a street. DOLCE distinguishes physical and temporal qualities, which pertain to endurants and perdurants, respectively. **Roles** are played by endurants. For example, a physical object may play the role "observed object", but it may also play the role, e.g., of an "operation site" or of a "target".

To exemplify the importance of such ontological precision: in (O&M), some vital concepts are under-specified or ambiguously defined. For example, "observed property" and "phenomenon" are defined vaguely and used more or less like synonyms. According to DOLCE, they would be a mixture of endurant, perdurant, and quality (see a detailed discussion in [19]). Similarly, "feature of interest" is not perceived as a role (which is done according to DOLCE), but instead as an endurant – although, quite clearly, being observed is not a characteristic property of an object. The Rhine is a river; will it become a different object because it is being observed? Such terminological inclarity is unproblematic when used amongst members of a closed community who know what is meant, but may cause problems when crossing community boundaries – e.g. during a disaster. That said, the GDO is not dogmatic in its alignment to DOLCE; we follow the DOLCE guidelines where sensible, and opt for pragmatic solutions in cases where a full solution would unnecessarily complicate matters.

The GDO is based on the design pattern depicted in Figure 2. That is, the ontology is built as a specialization of that pattern, extending the pattern's high-level categories with whole taxonomies, i.e., with hierarchies of more concrete categories, and instantiating the high-level relations with relations between such concrete categories. In what follows, we briefly explain the main aspects of the design.

At first glance, one sees that the pattern does not only cover sensor observations – **observable qualities** – but also **weather phenomenon, substance, geosphere region**, and **boundary of geosphere regions**. This enables *search by related terms*: rather than laboriously searching through a huge set of observable qualities, the user may select a related concept which pertains to the desired quality.[5] The advantage is that the

---

[5] The relation may be direct or indirect; hence the **has quality** and **has indirect quality** relations in Figure 2. To exemplify the difference: water (directly) has a temperature; in contrast, pressure is not a property of the athmosphere, but is often (indirectly) associated with it.

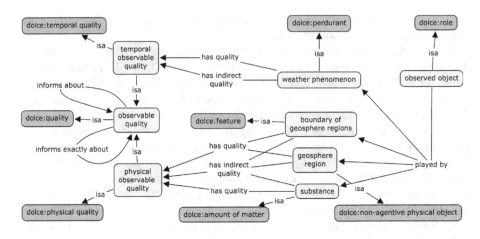

Fig. 2. The design pattern underlying the GDO (Geosensor Discovery Ontology), slightly simplified for presentation. Concepts inherited from DOLCE are marked by inscription and color.

taxonomies of related concepts tend to be much smaller than that of possible sensor observations. For example, for a non-expert user "wind direction" (or "water level") are probably much easier to find via "wind" (or "river") than via browsing the taxonomy of observable qualities. That said, browsing is of course also an option in our system.

In the GDO, **weather phenomenon** captures things such as rain shower, wind, fog; **substance** is orientated at chemical terminology, distinguishing between pure substances and blended subtances, covering things such as oxygen and nitratemonoxide (pure substances), and salt water (a mixture of substances); **geosphere region** covers things such as athmosphere, ground, body of water; **boundary of geosphere regions** covers things such as earth surface, water surface. If needed, these 4 top-level categories can easily be augmented by additional ones. One simply adds the new categories, classifies them according to DOLCE, and gives them the **played by** relation to **observed object** – which is defined as a role, c.f. the above discussion.

In accordance with DOLCE, observable qualities are distinguished into temporal ones (e.g. speed, flow rate) and physical ones (e.g. temperature, distance). Another aspect worth noting is that observable qualities may be related – one quality **informs about** another – or even equivalent – one quality **informs exactly about** another. An example of the former is fog density, which informs about range of sight. An example of the latter are the two ways of observing wind direction: *from where* vs. *whereto*.

4.3   Semantic Annotation

As stated, our semantic annotations are simple, in order to ensure practicality for organizations such as fire brigades. The precise form of the annotations is as follows:

Definition 1. *Assume that s is a SOS service. A service description of s is any set D that contains the URL of s as well as a* semantic annotation α *of s, defined as follows.*

Assume that $OP(s) = \{op_1, \ldots, op_k\}$ is the set of observed properties supported by s, across offerings, and assume that OQ is the set of concepts in the GDO that are sub-concepts of **observable quality***. Then a semantic annotation of s is a partial function $\alpha : OP(s) \mapsto OQ$.*

Sub-concept here refers to the taxonomic structure of the GDO: concept c_1 is a sub-concept of concept c_2 iff c_1 lies below c_2 (directly or indirectly) in the tree of concepts. In practice, and in our prototype, of course the form of the service descriptions (i.e., the precise set of attributes stored for each service) is fixed. What that form is – other than that it complies with Definition 1 – is not important to this work. Note that α is a partial function, hence allowing the annotation to be incomplete. This allows to register a service without giving it a full semantic annotation. In order to use a particular output (a particular observed property) of a service with our architecture, that output must be annotated, i.e., be in the domain of the annotation function α.

Each observed property is characterized by a single concept of the GDO. This is appropriate because it complies well with the intended meaning of the SOS specification: each sensor output corresponds to one atomic category of possible observations. It is important to note that such a simple correspondence would *not* be valid for more complex OGC services. For example, it would make no sense to restrict the annotation of a WFS service to a single concept in an ontology: since WFS services are databases that may contain a whole variety of data, a description of their data content would definitely need to be some sort of combination of concepts (see also [15]). From a Semantic Web perspective, ours is a classical example of a light-weight approach, c.f. Section 2.2.

In our architecture, the simple semantic annotations as per Definition 1 suffice to conveniently discover and, where needed, replace SOS services (details follow in the next sub-sections). Creating the annotations can, obviously, be supported in a straightforward manner using classical GUI paradigms. Figure 3 shows a screenshot of our implemented tool, in a situation corresponding to use case (C) of Section 3, i.e., annotation of air pollutant concentration measurements with concepts from the ontology.

As can be seen in Figure 3, the WSR GUI contains a tab for annotating sensor services. The WSR displays the service's observed properties, as well as any α assignments that have already been made. In a separate part of the window ("Konzepte"), the ontology is displayed. One can search concepts in the ontology via several options that will be detailed in the next section, when we describe how to create discovery queries. Once the desired concept is found, one simply drags it onto the corresponding observed property – in Figure 3, the concept "Lufttemperatur" is dragged onto the output property "airtemperature". The new assignment is stored in the service's annotation α. If the output was already assigned previously, then that assignment is over-written.

Clearly, this annotation process requires no more expertise than a basic familiarity with computers, as well as some familiarity with SOS service observations and with the GDO. It is realistic to assume that such expertise will be available, or easy to create, within the relevant organizations and their partners.

4.4 Sensor Discovery

As is common in semantic service discovery, c.f. Section 2.2, the discovery is formulated as a process of matching the available services against a discovery query. In our

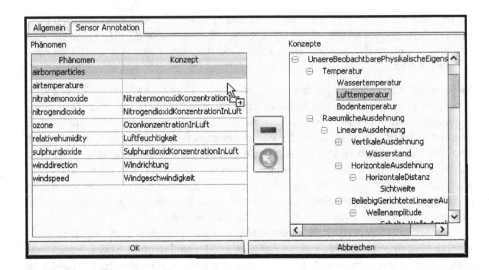

Fig. 3. A screen shot of our GUI for creating semantic annotations. Since our tool is built in cooperation with (and for the use of) German disaster defence organizations, the inscriptions are in German; explanations are in the text.

approach, the semantic annotations serve for terminology mediation, and for allowing *indirect* matches. The latter enables the user to find the desired services via intuitively related terms, rather than having to laboriously search for the actual technical term.

Service descriptions and the semantic annotations they contain were defined already in Definition 1. Discovery queries and matches are defined as follows:

Definition 2. *Assume that $\mathcal{CO}$ is the set of all concepts in the GDO. A semantic discovery query sQ is a subset $sQ \subseteq \mathcal{CO}$. Assume that D is the description of a service s, that $OP(s) = \{op_1, \ldots, op_k\}$ is the set of observed properties supported by s, and that $\alpha \in D$ is the semantic annotation of s. Then sQ and s match in op_i iff op_i is in the domain of α, $\alpha(op_i) = c_0$, and there exists $q_0 \in sQ$ such that q_0 is connected to c_0. The latter notion is defined inductively as follows:*

(1) Every $c \in \mathcal{CO}$ is connected to itself.
(2) If the GDO contains a relation with domain $c_1 \in \mathcal{CO}$ and range $c_2 \in \mathcal{CO}$, then c_1 is connected to c_2.
(3) If $c_1 \in \mathcal{CO}$ is a super-concept of $c_2 \in \mathcal{CO}$, then c_1 is connected to c_2.
(4) If $c_1 \in \mathcal{CO}$ is connected to $c_2 \in \mathcal{CO}$, and c_2 is connected to $c_3 \in \mathcal{CO}$, then c_1 is connected to c_3.

In words, a discovery query is just some collection of terms from the ontology. What the discovery does is to look for services s whose annotation contains a term c_0 which one of the query terms (namely q_0 in the definition) is "connected" to. All these services s – along with the relevant observation op_i and ontology term c_0 – are returned, provided the spatial and temporal aspects match as well (see below).

Connected in Definition 2 refers to a combination of relations in, and taxonomic structure of, the GDO. It is best understood as defining a set of possible paths through the ontology. Item (1) in Definition 2 says that empty paths are allowed: a query concept q is, of course, relevant to itself. Item (2) says that a path may follow a relation between two concepts c_1 and c_2 – if c_1 is relevant to the query, then c_2 is as well because c_1 relates to c_2. For example, c_1 may be the concept **river**, the relation may be **has quality**, and c_2 may be **water level**; c.f. use case (A) of Section 3. Item (3) in Definition 2 says that a path may go downwards in the taxonomy, i.e., go from c_1 to c_2 if c_1 lies above c_2 in the taxonomy. This is so because, if c_1 is relevant to the query and c_2 is a special case of c_1, then clearly c_2 is relevant to the query as well. For example, the query concept may be **body of water**, which is a super-concept of **river**, from which by item (2) we may get to **water level**. Item (4) states transitivity, a technical vehicle for expressing concisely whether or not there exists a path between two concepts.

Items (1)–(4) in Definition 2 are implemented in a straightforward way using F-Logic *rules*. Such a rule takes the form *rule-head* $\Leftarrow$ *rule-body*, meaning that truth of the rule body (right hand side) implies truth of the rule head (left hand side). Rule head and body are composed of F-Logic atoms. Item (4), e.g., is implemented by the rule $\forall$X,Y,Z `connected(X,Z) <- connected(X,Y) AND connected(Y,Z)`. While one could of course implement items (1)–(4) "by hand", the F-Logic implementation is efficient, and has the advantage of full flexibility: our approach and implementation can be trivially adapted to extended or modified matching methods, as long as the matching is expressible within the realm of F-Logic.

The above clarifies the *semantic* part of the discovery. On top of that, we need to specify the desired geographical region and time points. Consequently, a discovery query Q consists of a semantic discovery query sQ in combination with a bounding box bb and a time interval ti, both defined in the usual way. An observed property op_i of a service matches a query Q iff it matches sQ according to Definition 2, *and* the bounding box of the corresponding offering has a non-empty intersection with bb, *and* the time interval of the offering has a non-empty intersection with ti.[6]

Having clarified the inner workings of discovery, the important question remains how that functionality interfaces with the user. *How do non-experts such as fire brigade officers, acting under great stress, create discovery queries?* Given that our queries are combinations of standard constructs and very light-weight semantics, such query creation is quite feasible. Figure 4 shows the relevant screen shots for illustration.

We do not show a screenshot for specifying the bounding box and time interval because these interactions are obvious. The bounding box is specified within the GIS Plugin via marking a rectangle on the map. The time interval is specified via a time line with lower and upper bounds, shown at the bottom of the windows in Figure 4 (in the windows, the right-hand part of the interval has been selected). The core part of query creation consists of finding the desired set sQ of terms from the ontology. The WSR GUI offers three options: *text search*, *browsing*, and *following relations*. The first two facilities are illustrated in Figure 4 left-hand side, the third one is shown in Figure 4

---

[6] One can rank the services depending on the match quality. In our implementation, the ranking is a combination of the distance (path length) between the relevant query and annotation terms (q_0 respectively c_0), as well as the size of the intersections with bb respectively ti.

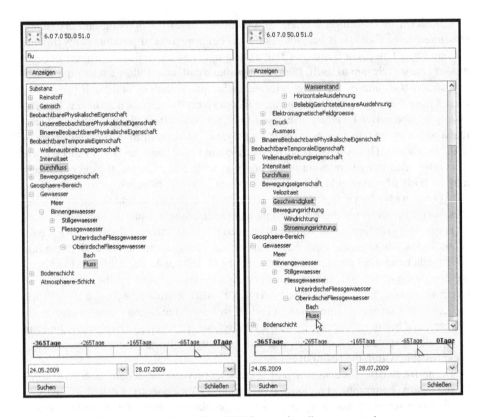

Fig. 4. Screen shots of our GUI for creating discovery queries

right hand side. On the left, the user has entered the text "Flu", which string-matches with "Fluss" (**river**); the GDO taxonomy tree is opened, highlighting that concept. Alternatively, the user could choose to browse for **river**, which would be done via clicking downwards in the taxonomy tree shown below "Geosphaere-Bereich" (**geosphere region**). On the right, the user wishes to give the precise phenomenon for the query, and chooses to look at the terms related to "Fluss" (**river**). This is done by a double-click on that concept. All related concepts, among them the desired "Wasserstand" (**water level**) are shown and highlighted.

Note how this form of discovery addresses problems (I) and (II) described in the introduction. Problem (I) – mediation is required between the terminology of the user and that of the Web service design – does not occur in Figure 4 because the required mediation has already been done at the point these interactions happen. The translation of terms is stored in the semantic annotations, and from the point of view of the end-user (who is likely to be different from the person doing the annotations) there *is* only one terminology. As for problem (II) – the user may not even know a technical term for the observed property she is looking for – this is addressed by the option to follow relations (Figure 4 (b)), and by the option to not even search for the actual phenomenon by hand but instead leave it up to indirect discovery (c.f. Definition 2) to make the connections.

Once the completed discovery query has been sent to the WSR, all matching services are returned. The user may simply select all these services, or, in case the query was for more general ontology terms, he/she may select a subset. To help with the latter, the WSR GUI offers the option to display, for each service, the actual observations (the ontology terms annotated at the service) that match the query.[7]

4.5 Sensor Fusion and Replacement

Our architecture also serves to fuse data from different sensor services (c.f. problem (III) from the introduction), and to replace damaged sensors through appropriate other sensors (c.f. problem (IV) from the introduction). This is realized by the Joint Sensor Engine (JSE). After the user has selected sensor services in the WSR GUI and dropped them into the GIS GUI, the GIS GUI sends a request to the JSE. The JSE retrieves the data from the SOSs, and transforms these as necessary. Afterwards, new observation layers are added to the map displaying the features of interest (FOI) as well as the actual sensor values. We now describe these functionalities in more detail. We ignore the case where the user selects only a single service in the WSR GUI. Obviously, this is simpler to handle than the more general case where several services are selected.

After the user has dropped the services onto the GIS GUI, a sensor request is created and sent to the JSE. This request includes the endpoints of the sensor services, a layer id, the observed properties, the sensor IDs, and a temporal and spatial extent. The JSE translates the sensor request into service-specific SOS requests, and calls the services accordingly. The SOS responses are then merged as follows.

First, depending how data is distributed over several SOS instances, there may be redundant data provided by more than one instance. For example, in our use case (A) from Section 3, two sensors for data "upstream of Cologne" and "downstream of Cologne" might duplicate the data for Cologne itself. The JSE checks whether such duplicates occur, by comparing the relevant concepts of the GDO. If the observed properties (i.e., the annotated concepts) are the same, and that is also the case for the FOIs and the time-stamps of the data, then only one of the duplicate values is considered.[8]

Second, data transformation may be necessary. Trivially, this is the case for units of measurement, which need to be normalized to the style of presentation used in the GIS GUI. This is done via standard techniques. The more interesting case is that of sensors which measure equivalent observable qualities, such as wind direction *from where* vs. *whereto*. Note that this is an important issue for crisis team work because, to correctly interpret such data, without IT support one needs to be aware of rather subtle context information – e.g. wind direction is interpreted differently in Germany and the Netherlands, so one would need to take the respective location of the service into consideration. The GDO resolves this issue via the aforementioned **informs exactly about** relation, c.f. Section 4.2. By virtue of the semantic annotations, the JSE knows that the observations are different; by virtue of the **informs exactly about** relation, the

---

[7] More advanced support may be possible relating to, e.g., quality-of-service parameters of the services. This is a direction for future work.

[8] Note here that the GDO is required for being able to do so: duplicate detection via sensor IDs is not possible because those IDs are not maintained globally, i.e., across SOS services.

JSE knows that they are equivalent. That said, our solution is preliminary in that the GDO does not state how to actually transform measurements of these observations into one another. To state this in the GDO, one would need to include arithmetic terms in the ontology. This is not possible in either of OWL or F-Logic. Our current implementation simply hard-codes this arithmetic into the JSE. A more flexible solution, e.g. via stating the arithmetics within ontology comments, is a topic for future work.

The JSE monitors service invocations, and automatically replaces a service if the monitoring concludes that the service is not functional anymore. We explain below exactly when that conclusion is made. First, we define what sensor replacements are:

Definition 3. *Assume that s is a service, that $OP(s) = \{op_1, \ldots, op_k\}$ is the set of observed properties supported by s, that D is the description of s, and that $\alpha \in D$ is the semantic annotation of s. Assume similar notations for another service s'. Then s' can replace s in op_i by op'_j iff: op_i is in the domain of α, $\alpha(op_i) = c_0$; op'_j is in the domain of α', $\alpha'(op'_j) = c'_0$; and c'_0 can replace c_0. The latter notion is defined inductively as follows:*

(1) Every $c \in CO$ can replace itself.
(2) If $c_1 \in CO$ is a sub-concept of $c_2 \in CO$, then c_1 can replace c_2.
*(3) If c_1 **informs exactly about** c_2 according to the GDO, then c_1 can replace c_2.*

Note that, as before, this definition covers only the semantic part of the replacement. In addition to the conditions stated, we require that the respective FOIs are identical, and that the respective time stamp of s' is at least as recent as the last valid measurement provided by s. Definition 3 should be largely self-explanatory. Item (1) is obvious, item (2) says that we can replace a sensor with a more specialized sensor, and item (3) states that we can replace a sensor with a sensor providing an equivalent observation. These items may be combined in an arbitrary fashion. For illustration of item (3), re-consider the wind direction example mentioned above. An example where item (2) is relevant is that were both s and s' measure the speed of a river, and op_i is annotated with **velocity** while op'_j is annotated with **stream velocity**. To illustrate item (1), consider use case (B) from Section 3, where water level sensors may require replacement.

To perform a replacement, the JSE contacts the WSR with a discovery query that contains the URL of s, as well as the desired observation op_i (if more than one op_i are needed, several queries are posed). The WSR returns the suitable replacements s' and op'_j as per the above. The replacement is triggered iff monitoring detects one of the following situations: the service does not respond; an error occurs; the answering time exceeds a given time interval; the observation values provided by a specific sensor are empty or outside a given interval.

5   Related Work

There are several projects in which OGC SWE services have been applied to risk monitoring and disaster management, e.g. [11,20]. In difference to our work, these projects focus on service architectures and SWE protocols for data exchange and fusion without any formalized knowledge.

There is some previous work on creating ontologies in the context of SOS services, most importantly the SWEET project[9] which has developed ontologies [21] that cover a broad spectrum of GIS terminology. The GDO models the SoKNOS-relevant subset of observable qualities defined in the CF Metadata and in SWEET. The ontology structure of SWEET was examined closely, and some relevant approaches were adapted to suit DOLCE. Overall, the GDO is more specialized than SWEET, and more suitable for our application; it is distinguished through its conformity with DOLCE.

Semantic discovery of OGC services has previously been investigated in the following three works. [15] design a Desciption Logics based approach to discovery of WFS services, and [14] design a 1st-order approach to WPS service discovery. Recent work has developed a more light-weight logic programming based approach to WPS discovery [4]. Although our approach uses similar machinery (F-Logic), there is no technical or conceptual relation between the two works.[10] Recent work [9] is similar to ours in that it also addresses semantic annotation of SOS services. However, the intentions, and consequently the employed methods, are very different. Whereas we aim at quick discovery and fusion of sensors in situations of great stress, [9] aim at a deep analysis of sensor *data*, automatically identifying phenomena such as blizzards from the sensor output. Thus, in stark difference to our light-weight annotation of SOS descriptions, [9] use more heavy-weight annotation and reasoning about data *content*. Finally, in [10] a method is proposed for linking Geosensor network data and ontologies. In difference to our work, the focus of [10] is mostly on the generation of annotations, the main contribution being an implementation of such methods within the Protégé ontology editor; also, the application domain is different, namely transportation.

6   Conclusion

We have presented an architecture for flexible discovery and integration of SOS services, based on light-weight semantic annotations. The annotations are sufficiently easy to create for end-user acceptance, while at the same time they provide significant added value through the ease of finding suitable sensors, and the ease of fusing their data and dealing with service failure. Hence ours appears to be a good compromise between the power of semantic annotations and the difficulty of creating and maintaining them.

An evaluation by real fire brigade men has largely confirmed this view. Three groups of men ranked the discovery functionalities – text search, browsing, linked concepts, indirect discovery – with school grades. All grades were among the best 2 grades available, and top grades were given 5 times. The men expressed the view that such a tool would be useful for crisis team work. They were especially enthusiastic about indirect discovery (discovery via related terms) because, under the stress of a crisis, it will often happen that crisis team members don't immediately recall the correct technical terms.

Of course, our architecture is far from perfect and several issues have been left unaddressed as yet. Some important ones regard the selection of services, once a set has

---

[9] Semantic Web for Earth and Environmental Terminology, http://sweet.jpl.nasa.gov/index.html

[10] [4] focus exclusively on formulating the dependencies between inputs and outputs of a service – an issue which does not even arise for SOS services. While we match through a notion of paths through the ontology, [4] use query containment.

been discovered. Our current ranking methods are fairly primitive. A tool for quickly comparing services, i.e., showing at one glance their most relevant strong/weak aspects, would be desirable. Also, depending on the level of user acceptance of creating more complex annotations, techniques such as presented in [9] (c.f. Section 5) may be quite useful for automatically issueing warnings regarding potentially dangerous events.

Acknowledgments

Part of this work was supported by the German Federal Ministry of Education and Research (BMBF) under grant number 01—S07009 and the SAP AG within the context of the SoKNOS project.

References

1. Botts, M.: OpenGIS Sensor Model Language (SensorML) implementation specification, version 1.0.0. Tech. Rep. 07-00, Open Geospatial Consortium (2007)
2. Botts, M., Percivall, G., Reed, C., Davidson, J.: OGC white paper - OGC Sensor Web Enablement: Overview and high level architecture. Tech. Rep. 07-165, Open Geospatial Consortium (2007)
3. Cox, S.: Observations and Measurements - part 1- observation schema version 1.0. Tech. Rep. 07-022r1, Open Geospatial Consortium (2007)
4. Fitzner, D., Hoffmann, J., Klien, E.: Functional description of geoprocessing services as conjunctive datalog queries. Geoinformatica (2009) (currently under review)
5. Gangemi, A., Guarino, N., Masolo, C., Oltramari, A.: Sweetening WordNet with DOLCE. AI Magazine 24(3), 13–24 (2003)
6. Gangemi, A., Mika, P.: Understanding the Semantic Web through Descriptions and Situations. In: Meersman, R., Tari, Z., Schmidt, D.C. (eds.) CoopIS 2003, DOA 2003, and ODBASE 2003. LNCS, vol. 2888, pp. 689–706. Springer, Heidelberg (2003)
7. Gruber, T.: A translation approach to portable ontology specifications. Knowledge Acquisition 5(2), 199–220 (1993)
8. Hendler, J.: On beyond ontology: What's next for the semantic web? In: Fensel, D., Sycara, K., Mylopoulos, J. (eds.) ISWC 2003. LNCS, vol. 2870. Springer, Heidelberg (2003)
9. Henson, C.A., Pschorr, J.K., Sheth, A.P., Thirunarayan, K.: SemSOS: Semantic Sensor Observation Service. In: International Symposium on Collaborative Technologies and Systems, CTS 2009 (2009)
10. Hornsby, K., King, K.: Linking geosensor network data and ontologies to support transportation modeling. In: Nittel, S., Labrinidis, A., Stefanidis, A. (eds.) GSN 2006. LNCS, vol. 4540, pp. 191–209. Springer, Heidelberg (2008)
11. Jirka, S., Bröring, A., Stasch, C.: Applying OGC Sensor Web Enablement to Risk Monitoring and Disaster Management. In: GSDI 11 World Conference, Rotterdam, Netherlands (June 2009)
12. Kifer, M., Lausen, G., Wu, J.: Logical Foundations of Object-Oriented and Frame-Based Languages. J. ACM 42(4), 741–843 (1995)
13. Li, L., Horrocks, I.: A software framework for matchmaking based on semantic web technology. In: 12th International Conference on the World Wide Web, WWW 2003 (2003)
14. Lutz, M.: Ontology-based descriptions for semantic discovery and composition of geoprocessing services. Geoinformatica (2006)

15. Lutz, M., Klien, E.: Ontology-based retrieval of geographic information. International Journal of Geographic Information Science 20(3), 233–260 (2005)
16. Masolo, C., Guarino, N., Oltramari, A., Shneider, L.: The Wonder Web library of foundational ontologies. Tech. rep. (2003)
17. McGuinness, D.L., van Harmelen, F.: Web Ontology Language (OWL) Overview, W3C Recommendation (February 2004), http://www.w3.org/TR/owl-features/
18. Na, Arthur Priest, M.: Sensor Observation Service - implementation specification version 1.0. Tech. Rep. 06-009r6, Open Geospatial Consortium (2007)
19. Probst, F.: Semantic Reference Systems for Observations and Measurements. PhD Dissertation (2007)
20. Raape, U., Teßmann, S., Wytzisk, A., Steinmetz, T., Wnuk, M., Hunold, M., Strobl, C., Stasch, C., Walkowski, A.C., Meyer, O., Jirka, S.: Decision support for tsunami early warning in indonesia: The role of standards. In: Cartography and Geoinformatics for Early Warning and Emergency Management (2009)
21. Raskin, R., Pan, M.: Knowledge representation in the semantic web for earth and environmental terminology (SWEET). Computers and Geosciences 31(9), 1119–1125 (2005)
22. Vemmer, T.: The Management of Mass Casualty Incidends in Germany - From Ramstein to Eschede. BoD (2004)

A Spatial User Similarity Measure for Geographic Recommender Systems

Christian Matyas and Christoph Schlieder

Laboratory for Semantic Information Technology, University of Bamberg,
Feldkirchenstrasse 21, 96052 Bamberg, Germany
{christian.matyas,christoph.schlieder}@uni-bamberg.de
http://www.kinf.wiai.uni-bamberg.de/

Abstract. Recommender systems solve an information filtering task. They suggest data objects that seem likely to be relevant to the user based upon previous choices that this user has made. A geographic recommender system recommends items from a library of georeferenced objects such as photographs of touristic sites. A widely-used approach to recommending consists in suggesting the most popular items within the user community. However, these approaches are not able to handle individual differences between users. We ask how to identify less popular geographic objects that are nevertheless of interest to a specific user. Our approach is based on user-based collaborative filtering in conjunction with an prototypical model of geographic places (heatmaps). We discuss four different measures of similarity between users that take into account the spatial semantic derived from the spatial behavior of a user community. We illustrate the method with a real-world use case: recommendations of georeferenced photographs from the public website Panoramio. The evaluation shows that our approach achieves a better recall and precision for the first ten items than recommendations based on the most popular geographic items.

Keywords: recommendation, personalization, geospatial services.

1 Introduction

With the wide acceptance of the social web idea, building collaborative georefenced data libraries has became popular in the recent years. Public websites, like flickr[1] and panoramio[2], offer huge georeferenced datasets of user generated images that depict buildings or landscapes. New location-based social networks like Citysense or BrightKite enable people to publish their current location as a point based feature enhanced by additional information, like multi-media or ratings for that position. By accessing open software platforms, like Google Latitude or Yahoos Fire Eagle, almost any application can attain location awareness.

---

[1] Flickr: www.flickr.com: 76,711,981 images (june 2009).

[2] Panoramio: www.panoramio.com. 13,005,132 images (june 2009).

K. Janowicz, M. Raubal, and S. Levashkin (Eds.): GeoS 2009, LNCS 5892, pp. 122–139, 2009.

The integration of personal location awareness has achieved general acceptance in the popular iPhone and Android capable devices.

As a consequence, an ever increasing number of mobile web users do not just create data (e.g. photographs) but georeferenced data (e.g. photographs with GPS footprints). Very often, this happens without these users being aware of having created volunteered geographic information [1]. The enormous quantity of data that can be accessed via spatial features makes information filtering a central challenge of the emerging Geospatial Web [2]. Searching and recommending constitute the dominant information filtering paradigms. Whereas a search addresses users with information needs that can be stated explicitly in the form of key words to a search engine, recommendations are appropriate when information need is conveyed implicitly by personal preferences which are difficult to express verbally. A recommender system predicts what data objects could be of interest to a user. The prediction is based upon previous decisions of the user as well as upon decisions that others from the user community have taken.

A geographic recommender suggests items from a library of georeferenced objects [3]. As an illustrative example, imagine the task of recommending a set of touristic photographs. We could, for instance, request six images depicting the city of Bamberg, Germany that match our personal preferences for content and style. These images are then used to generate our personalized patchwork postcard of the city (figure 1). Whether we select 6 images for a patch work postcard or 50 images for a slide show, in both cases we express a pictorial conceptualization of a geographic place that includes certain sights and excludes others. Two questions are of immediate interest to the design of a geographic image recommender system: (1) What is the typical pictorial conceptualization of a given place, in other words, which images are most often used to illustrate that place? (2) What is the personalized conceptualization of a place, that is, the selection of images that a specific user would prefer to see? As the first question has been addressed in detail by Schlieder and Matyas [4], our paper focuses on the second.

One empirical finding from our study of typical conceptualizations was that the decision of a user to publish a touristic photograph in a web collection is

Fig. 1. Use case for a geographic recommender: patchwork postcards

not impeded by the fact that other users have already published similar images. Case in point, users have independently uploaded 1,500 photos of the Spire of Dublin, or in a more extreme case, around 25,000 pictures of the Eiffel Tower in Paris on flickr. As national landmarks, this gives a rough idea of the popularity of them. We explore the idea of popularity by assuming each of these uploads can be interpreted as a vote for the geographic place. Places can then be ranked by their popularity forming a hit list of the most prominent instances. We call this ranked list a *heatmap*.

The simplest type of recommendation is generated by taking into account the decisions of the community while ignoring decisions of the user for which the recommendation is generated. A recommendation that suggests images depicting the most popular sites can easily be generated from the heatmap. The problem with this approach is obviously that individual preferences may differ considerably from the aggregated preferences of the community. The *collaborative filtering* approach to recommending pioneered by Resnick et al. [5] solves the problem by comparing users on the basis of the decisions they have taken following the idea that *"people who agreed in the past, are likely to agree again"*. We transfer this principle to the spatial domain and refine the assumption so that:

> *"People who agreed on the qualities of one geographic region, are likely to agree on the qualities of other geographic regions too."*

Collaborative filtering allows us to use the user's feedback to recommend new and undiscovered items. In the example of the patchwork postcards, a user could query a geographic recommender for a personal selection of six images for Dublin although he has never been there (figure 1). Images that the user has chosen in the past to illustrate another city, say, Bamberg, are used as information source or feedback to the recommender. As far as we know the geographic footprint of a data object, in our case, the geographic coordinates associated with the image, consitute a previously unexplored source of feedback for a recommender system. The paper makes three major contributions:

1. Based on previous work [4] we extend the idea of frequency-based analysis of collaborative geographic data to encompass the analysis of a spatial partonomy. We introduce a measurement that weights the nodes in this partonomy to support user similarity (section 3).
2. We show how a user-based collaborative filtering approach can be used in conjunction with the weighted partonomy. We discuss four different measurements that estimate the user similarity on different levels of the partonomy. Based on the user similarities we generate a personal recommendation of geographic objects (section 4).
3. We investigate the performance of the four measurements in an example situation of a geographic recommender system, compared to a impersonal recommendation based on popularity. We evaluate their precision and recall at rank 10 by drawing on over 30.000 georeferenced images from the public image gallery Panoramio (section 5).

2   Related Work

Several researchers extract information about places based on geographic metadata. Girardin et al. [6] tried to understand touristic dynamics by visualizing the data in different temporal contexts. Of more interest towards geographic modeling, recent work uses spatial and temporal clustering to extract typical places and semantic information of different regions [7][8]. Most of these works are based on the use of simple feature detection techniques like SIFT point matching in the images [9][10]. A common ambition is to improve the users experience while browsing massive georeferenced image sets. We employ geographic modeling in our recommendation approach. We use a two step method, deriving the geographic model for use in a user-based collaborative filter. Burke [11] refers this as a hybrid meta-level recommendation because of the combination of two different recommendation methods. Besides our chosen modeling the method can be supported by any geographic model previously cited.

Classical Recommender systems filter the information based on the user that demands the information [11]. Resnick et al. [5] introduced the first successful recommendation system based on user ratings for newspaper articles, called Grouplens. The approach has since been adopted by many applications, but they all still depend on an explicit rating of the objects (e.g. [12], [13]). They also retain a common working principle of user-based filtering. The user-rating vectors of individuals allow comparisons between users in order to find people that agreed (or consistently disagreed) in the past. The ratings of these users are combined to give a predictive rating. We find many algorithmic approaches based on ratings trying to improve the results of Resnick (e.g. [14] [12] [15] [16]).

Suggesting the n most interesting items for a user with the highest amount of diversity is one challenge. While we concentrate on the geographic metadata to surmount this problem, other semantic driven approaches can also be found in the literature [17] [16]. Including the user-based approaches, item-based collaborative filters have become very common in modern eCommerce applications, as seen on Amazon [18]. In contrast to the user-based collaborative filtering this technique looks for relationships between the items, ignoring the measure of user similarity. This approach scales for huge unstructured data but fails to capture the users' differences, which is our main goal.

Similarity has always been a major topic in many scientific fields, most of them based on the article of the psychologist Tversky [19] about similarity in general. Geographic similarity based on ontology of places has primarily been used to support information retrieval [20] rather than recommender systems. Much research has been dedicated to discover appropriate ways to measure the similarity of spatial feature sets. Rodriguez and Egenhofer [21] and Schwering [22] gave a good survey of the major results, whereas recent work can be found in Janowicz et al. [23]. The feature set approach to defining similarity is often adopted when cognitive adequacy is important. There is little work, however, on the relationship between feature similarity and user similarity. Information theoretical approaches to concept similarity have been studied mainly outside geospatial applications in connection with taxonomic reasoning [24].

The idea of using recommendations in a geographic environment is not totally new. Notably, point features were used in recent work to generate a more relevant result when the user is moving in a mobile environment. The term location-based recommendation is used in the literature to classify this kind of work. Bae-Hee et al. [25] takes the current GPS position to introduce a location context in addition to the personal context and the environmental context for the generation of the final recommendation. The rating for this location context is the inverse ranking of the n-nearest items, which makes the final results more dependent on the current position of the user. Their approach uses an additional weight to reorganize the final results whereas we base our whole approach on geographic input data. Horozov et al. [26] uses a circular buffer around the current position of the user in order to reduce the possible recommendation candidates. They also evaluate the influence of demographic information, like sharing the same neighborhood, on the precision of the recommendation results. Inspired by their idea of using additional geographic metadata to improve results, we identify complex data structures in different geographic regions (see section 3). This enables us to ask for recommendations like the *ten typical images from Bamberg*. Using a user-based collaborative approach we can also answer queries like: *If I would go to Bamberg, which are the ten images that I would have made?*

3   Collaborative Geographic Data

The applications mentioned in the introduction have to deal with volunteered geographic data that we refer to as collaborative geographic data. That means a user contributes his geographic objects to the public library and is not hindered or influenced by existing objects. The users are just providing their impression of the space in the form of a collection of point-based features (GPS positions). The decisions themselves can be linked to any kind of content like textual information, multimedia objects or ratings. In our use case we consider images linked to a specific position. We now encapsulate the content itself and focus our attention on the spatial choice expressed by this link. We use the term choice for situations in which the user could decide to act differently. In that sense, the user chooses to upload a particular image. We do not imply that choices are the result of explicit reasoning processes or associated with the conscious experience of taking a decision. Users usually have to physically move to reach a specific position in order to obtain the object they are providing for the system. Schlieder and Matyas [4], for example, identified two different spatial choices that a photographer is taking while shooting an image from a certain position. Firstly, the photographer has to move to a suitable location, from which he shots a photo. Secondly, he chooses the object(s) he wants to capture from his current position. Bae-Hee et al. [25] and Horozov et al. [26] also realized necessity of spatial decisions while choosing a restaurant for dining. In general, we are talking about decisions that have been made by a user and that can be represented by a specific geographic position, which we call a spatial decision or spatial choice. Users make spatial decisions everywhere and some of them will collide at the same location. The problem that arises from this observation relates to the need to identify identical spatial decisions in a semantically complex world. We simplify the situation by assuming two decisions as equal if they are relatively near one another. After reviewing various measures to compare spatial decisions of two users, we extend it with a frequency model of the geographic context. We discuss the effect of focusing on different spatial contexts and the need for a geographic separation of the space to enhance expressions of similarity.

We can express spatial decisions as a feature set $A = a_1, \ldots, a_m$ where a_i is a spatial choice made by the user. This feature set can now be compared to the choices of a second user which constitute another feature set $B = b_1, \ldots, b_n$. In general the cardinalities of both sets are not equal. Decisions of two users are three different subsets like illustrated in figure 2. Many similarity measurements have been proposed working on spatial feature sets. The most popular one is a variant of the Tversky measure [19] published by Rodriguez and Egenhofer [21] using the constraint $\alpha + \beta = 1$.

$$sim(A, B) = \frac{|A \cap B|}{|A \cap B| + \alpha |A \setminus B| + \beta |B \setminus A|} \tag{1}$$

Violating the constraint of the above condition and using $\alpha = \beta = 1$, the measure simplifies to a variation of Tversky, the Tanimoto measure [27]:

$$sim(A, B) = \frac{|A \cap B|}{|A \cup B|} \tag{2}$$

Tversky also made the observation that not every possible feature has the same effect on similarity. He calls this the diagnostic value of a feature. We argue that the diagnostic value is relative to a spatial decision's popularity in a user community. Our metrics of popularity of a spatial decision depends on the number of users and their decisions made at the same location. The function $decisions_n(u)$ returns the number of decisions made by user u at a specific position n. Note that we only consider locations where we find at least one decision, so that $decisions_n(u) > 1$ by definition. The popularity of that position calculates as follows, where U_n is the set of users that made decisions at the same position:

$$popularity(n) = \sum_{u \in U_n} (1 + \log decisions_n(u)) \tag{3}$$

In order to use the popularity as the a diagnostic value we make the following assumption: *People that share more lower ranked decisions are more similar than*

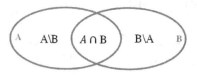

Fig. 2. Sets of spatial choices made by two users

people that share more higher ranked decisions. Figure 3a and 3b illustrates the impact of that assumption on the similarity. If we would weight the decisions equally the tanimoto measure would be same for both examples. $sim(A, B) = \frac{|A \cap B|}{|A \cup B|} = \frac{3}{6} = 0.5$. Depending on the weighting of the different ranks in the second case (figure 3b) would be much more similar because the value of the overlapping lower ranked decisions should be greater than in the first case (figure 3a). The problem of the ranking approach is the high dependency on the geographic context, which can hide regional differences. The larger we extend the geographic context, the smaller the relative significance of the individual ranking is. The most important issue for the similarity of two users is the relative ranking to an overall context. It is difficult to capture the differences of two users who only made spatial decisions in a city when considering the whole country.

The use of different levels of spatial separation offers the possibility to better model the user behavior. What we are looking for clearly separable environments where users usually make decisions. If we identify the typical environments in which users typically circulate, in order to make spatial decisions, we then reflect them in the partonomy as separate regions. For example tourists, a main source of images from public image galleries, commonly visit a few particular cities, resulting in a very high coverage of individual cities. In which case we profit from a representation of cities in our partonomy. There are a number of predefined partonomies, like the Nomenclature of Territorial Units for Statistics (NUTS), that can be used as background knowledge for defining these geographic contexts. NUTS results in a hierarchical partitioning of space where each level in the partonomy compromises a tesselation. The regions of a tesselation covers the complete space without overlap, see figure 4 for an example. The advantage of using a tessalation is also the complete coverage of all possible spatial decisions, which is advisable. We have not investigated overlapping regions yet, which would lead to slightly different results as some spatial decisions are considered multiple times.

We decided that a hierarchical partonomy gives as the necessary levels of granularity. This allows a broader interpretation of spatial decisions like mentioned before. We are also able to accumulate decisions made in lower layers to result in a more general idea of spatial decisions, like cities that a user visited instead of point based locations. The hierarchical partonomy can be described as a graph $G(N, E)$, where N defines a set of nodes and E a set of edges. The nodes $n \in N$ of the graph represent regions and each edge $e \in E$ represents a part-of

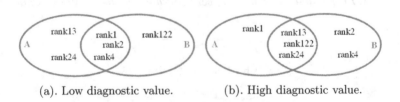

(a). Low diagnostic value.   (b). High diagnostic value.

Fig. 3. Different ranking combinations

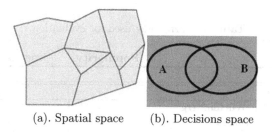

(a). Spatial space   (b). Decisions space

Fig. 4. Tesselation partitiong

relationship of two regions. The lowest level in our partonomy contains regions of cities, for example table 1. For the recognition of common spatial decisions we have to introduce another layer in our partonomy. This is due to the fact that even when two users visit the same location their GPS coordinates will show some slight differences. To compensate for this the lowest layer consists of clusters of point based features that permit recognition of the same spatial decision. We implemented a software application called the heatmapper that basically uses a geographic approach to cluster these features. We can also use the results of different approaches that are extracting places in such datasets (Ahern et al. [7], Snavely et al. [9]). Even an additional handmade modeling of clusters of spatial choices is imaginable. Whatever the case, if the model clusters all similar decisions, we can just choose any one of them to represent the cluster, e.g. a random image for a cluster of images. The regions occupied by the clusters, and not the points, are now the smallest geographic objects of interest and take place as leaves in the hierarchical tree. We introduced the term cluster of points of view (CPV) for a cluster of spatial decisions made while shooting images. Generally, we will talk about *clusters of spatial decisions (CSD)*.

For each node we calculate two different values in order to support our similarity measurements, it's popularity and derived from that the diagnostic value of that node. The popularity will be defined recursively: first we measure the popularity as described in equation 3 for every leaf, the rest of the nodes is defined as a sum of the popularity of its children.

$$popularity(n) = \sum_{c \in children(n)} popularity(c) \tag{4}$$

The second value we measure can be seen as the inverse of the popularity, as it reflects the diagnostic value of the node reflecting Tverskys notion of diagnosticity mentioned above. The weight $w(n)$ of a node reflects that certain choices occur more often or are considered more important than others, while the more common decisions will have a lower value than the more personal ones. As a supporting value we calculate the information content of a node by the formula, where $rp(n)$ is the relative popularity of a node in relation to the popularity of its siblings (siblings are nodes that share the same parent):

$$information(n) = -\log_2 rp(n) \tag{5}$$

Table 1. Partonomy used for evaluation

Region	Popularity	Images	Users
World			5766
Germany			2763
Baden-Würtemberg			1410
Stuttgart	2904.76	3471	808
Freiburg	1684.90	2047	505
Tübingen	544.14	638	171
Bavaria			1170
Munich	2649.25	2977	637
Nürnberg	138.72	195	44
Bamberg	1844.70	4692	261
Würzburg	1054.94	1209	298
Berlin			350
Berlin	2854.12	3258	350
Italy			2042
Toscana			1693
Pisa	747.14	793	494
Florence	3238.81	3988	876
Lucca	1328.80	1569	498
Lazio			508
Rome	2535.42	2858	430
Santa Marinella	44.04	49	24
Fiumicino	88.42	99	45
Aprilia	241.35	297	18
France			1461
North France			1461
Paris	3487.95	3784	598
Le Mans	315.35	390	89
Caen	624.69	691	182
Saint-Malo	1618.83	1735	658

This scales the relative popularity on a logarithmic scale and flatten big differences amongst the popularity values. The information content quantifies the value of information about a user's decision if that user participates in the corresponding node. We can say that a user participates in a node if one of his spatial choices is made inside the region represented by the node. It is obvious that a user participated in a node if he participated at least in one of the children nodes. The weight function $w(n)$ for a node n is measured as follows and scales the information content to a value between zero and one:

$$w(n) = \frac{information(n)}{\max_{s \in siblings(n)} information(c)} \tag{6}$$

The different participation patterns in sense of collections of spatial decisions for different spatial contexts can be evaluated to calculate the similarity between two users. We discuss different approaches in the next section. The tree structure allows us to exclude lower layers from consideration for a number of benefits. It reduces the complexity of the calculation at the cost of reducing the accuracy of the similarity. Additionally, we could find similarity otherwise not evident due to a lack of any common participation on the lowest level. Two users who visited the same city even if they have not participated in the same places in that city. Or as a more extreme example, users that have participated in the same country but not in the same cities. Each measurement should be able to differentiate between different levels of possible overlap of spatial decisions.

4   User-Based Collaborative Filtering

The initial task for this section is to generate a personal recommendation for users without explicit semantical information. In this sense we use a implementation of the prototype theory as stated by Rosch [28]. Every user has provided prototypes of different geographic contexts of the partonomy and they have been accumulated by the weighting of section 3 into a semantic model of the location. The ranked children of a node denote a typical conceptualization of that region. Using that ranking we can give a rather impersonal recommendation by returning the most popular decisions of this set of children. We will use this approach as a baseline for the evaluation in section 5. In order to give a more personal recommendation we base the calculations of a concept not on the whole community but on the most similar users to the initiator. In respect to Rosch, we use the prototypes of the users that shared the same experiences before to generate a concept of the region for that specific user group alone. This idea of using implicit user semantics can be seen in the implementation of a user-based collaborative filtering which we adopted for the recommendation of geographic objects.

Since user-based collaborative filtering was introduced by the Grouplens system (Resnick 1994) it always followed the same principles. Mandatory for the method is a constant feedback of the user. In the original work ratings about newspaper article were used as user feedback. Based on these ratings $r \in R$ the first step towards a recommendation is the identification of the most similar users. The Grouplens system used the Persons correlation factor of the user's u ratings vector $R_u = r_1, \ldots, r_n$ as similarity measurement. As a final step the opinions of the most similar users $N_u \subset U$ about a yet unrated item i are aggregated and used as a predicted rating r. The aggregation is basically done using the following formula, where $\overline{R_u}$ is the average rating of user u.

$$\widehat{r_{u,i}} = \overline{R_u} + \frac{\sum_{u \in N_u} (r_{u,i} - R_u) \cdot sim(u,u)}{\sum_{u \in N_u} |sim(u,u)|} \qquad (7)$$

We take the same basic steps but we use different similarity measurements based on the observations made in section 3 and an adapted aggregation approach. Both steps (finding a similar user and aggregating their experiences) will be described in the following. The first question we have to answer is: how similar are two users in respect to their spatial choices made in the weighted partonomy tree? We specify four possible measurements that are based on different levels of the derived semantic model.

1. **Single-layer feature similarity (SFS):** In a typical user based recommender the given feature vectors of each user I_U are measured against the feature vector of another user, using a correlation metrics like the cosine similarity or the Pearsons correlation factor, to find their similarity value. In our case we take the nodes of one level as features in a users feature vector (for example all CSDs). The values of the vector are the number of decisions a user made in each node. The similarity of two users ($u_a, u_b \in U$) is the comparison of these vectors using the cosine similarity. We choose cosine in order to compensate the differences in the number of images in one node, as we are more interested in the relative distribution of images on the different nodes.

$$sim_{SFS}(u_a, u_b) = \frac{I_{u_a} \cdot I_{u_b}}{|I_{u_a}| \cdot |I_{u_b}|} \tag{8}$$

2. **Two-layer feature similarity (TFS):** This measurement focuses on a single layer of the partonomy and calculates the cosine similarity for each node independently using the children of that node as feature set. The similarity in each node is used to scale the weight w(n) of the node, in the interval of [0,1]. The sum of the scaled node weights are then divided by the sum of the original weights. Nodes are weighted depending on their relative popularity. We only take nodes into account that both users have visited. $sim_{SFS}(u, u, n)$ is the cosine similarity restricted to children of node n as feature vectors.

$$sim_{TFS}(u_a, u_b) = \frac{\sum_{n \in cities} sim_{SFS}(u_a, u_b, n) \cdot w(n)}{\sum_{n \in cities} w(n)} \tag{9}$$

Generally, we can take any level in the partonomy graph and calculate the cosine similarity on the next lower level. We can calculate the feature similarity based on the cities to measure the similarity on the level of federal states as seen in table 1.

3. **Two-layer information similarity (TIS):** This measurement uses the same approach as the two layer similarity but is based on a similarity taking the information content (equation 5) of each CSD into account as introduced in section 3. The measure $sim_i(u_a, u_b, n)$ measures similarity based on the information value and a tanimoto theme in the context of one specific node n. Sets B_n and A_n are the set of participating children of the parent node n in relation to user u_a and user u_b.

$$sim_i(u_a, u_b, n) = \frac{\sum_{(c \in A_n \cap B_n)} information(c)}{\sum_{(c \in A_n \cup B_n)} information(c)} \tag{10}$$

$$sim_{TIS}(u_a, u_b) = \frac{\sum_{n \in cities} sim_i(u_a, u_b, n) \cdot w(n)}{\sum_{n \in cities} w(n)} \tag{11}$$

4. **Geographic coverage similarity (GCS):** This takes the common coverage on every level in the partonomy into account. As most users have some
common behavior at the higher levels of the partonomy this measurement
is relatively larger than the values of the other measurements. However we
are able to find a similarity even if the amount of common decisions in the
CSDs is low or not existing. $N_u, N_u \subset N$ are subsets of the nodes in the
partonomy graph user A respectively user B made a spatial decision.

$$sim_{GCS}(u_a, u_b) = \frac{\sum_{n \in N_{u_a} \cap N_{u_b}} w(n)}{\sum_{n \in N_{u_a} \cup N_{u_b}} w(n)} \qquad (12)$$

Basically, two users will be highly similar if they visited the same European
countries, within these countries chose similar regions, and within the regions
comparable cities.

Having established the similarity to other users we are able to finally calculate
a personal weighting of the nodes in the graph. In order to give each node a
personal weighting $w_{personal}$ we use the following binary function. $\gamma(u, n)$ equals
one if the user u participates in the node n and zero otherwise. $U(sim) \subset U$ are
the nearest neighbors of a user u.

$$w_{personal}(u, n) = \sum_{u_s \in U_{sim}} sim(u, u_s) \cdot \gamma(u_s, n) \qquad (13)$$

It is obvious that for every cluster in which none of the nearest neighbors participated, the personal weighting will accumulate to zero. The more of the nearest
neighbors a node has in common, the higher the value of that node will be. The
weight of the node will even rise faster if this neighbor is otherwise very similar
to the initiator.

This measurement is an adaption of the general aggregation approach
(equation 7), with some modifications. The original measure was scaled by the
sum of all similarities of the user. As we are only interested in the ranking of the
nodes we can ignore this factor, as it does not change the ordering. The differences to the average are exchanged with a the binary decision function $\gamma(u, n)$.
Figure 5a and 5b shows the impact of the personal weighting in relation to the
original weighting using the popularity of the nodes. The example is showing
the first 80 $CSDs$ in Bamberg that have been reweighted for one user. Figure 5a
shows the typical rank-popularity distribution based on the popularity measure
described in section 3. We see that it follows a power law typical for user generated content [29]. We also see is that many places are very prominent and
have a very high popularity while most places are observed by just some users.
Anderson [30] calls this a long tail distribution. Most of the items in the long
tail are only relevant for some people. The recommender system should be able
to speculate which elements of the long tail are relevant for the current user.
In figure 5b we see that some previously lower ranked clusters of the long tail
become much more prominent after the new personal weighting. An evaluation
will now have to prove how many of the items in the new top-n of the new
ranking correlate with the user preferences.

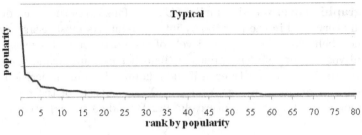

(a). Ranked by the popularity measurement.

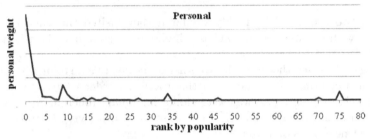

(b). Ranked by the popularity measurement, but weighted by the personal collaborative filtering method.

Fig. 5. First 80 clusters of Bamberg

5   Evaluation

In order to illustrate the full approach we demonstrate it on the use case of recommending geotagged images from a public collaborative image library. In June 2009 over 13 million images were accessible on Panoramio. Every image on Panoramio is geo-referenced using latitude and longitude information either from a GPS device or by self-positioning on a map interface. The dataset fulfills any constraints we discussed in section 3 for collaborative geographic data sets. As a result we are able to find multiple images for most of the tourist highlights all over the world (e.g. about 13,000 images of the Eiffel Tower and about 700 images from the Spire of Dublin). One advantage of panoramio is the focus of the collected data on having images of places ("Panoramio is different from other photo sharing sites because the photos illustrate places", as written in the help text on the site). This permits the suggestion that most user have a common motivation to upload images. Other sites would overcome diverse motivation by filtering specific images without real relation to its GPS position from consideration, like photos of families.

We previously showed that we can actually expect a power law pattern in respect to the popularity of the objects found in a specific region, take figure 5a as an example. Our aim is the recommendation of an image set of ten images of a specific city. The user selects a node in the partonomy the system ranks the children of this city node, in our case the calculated $CSDs$, using their popularity value. We are now able to give a first recommendation using the impersonal baseline approach, which is the Popularity (Top10) approach in figure 7a. This approach is later compared to the recommendation results. If our assumption holds true the recommendation results should correlate better with the user's actual decisions. The test was performed on a subset of images from panoramio, 33,947 images from 5,766 users in total. We identified 19 different cities in 3 different countries and fetched the images using the public API of panoramio. We choose the different cities based on the users of Bamberg, so that most users who made spatial choices in Bamberg also uploaded images in one of the other cities. This characteristic made them good test candidates for recommendations of Bamberg as we expect high overlap in lower levels of the partonomy among these users. The partonomy graph reads as follows (table 1), where the popularity and count of images are the sum of the lower nodes.

The evaluation uses a cross validation approach [31] which splits the available dataset in two separate non-overlapping datasets: a training set that is used to calculate a recommendation and a test set that is used for comparison. We repeatedly selected a user who uploaded at least 5 images for Bamberg as a test candidate and the training set consisted of the remaining decisions (see figure 6). For every test candidate, we first excluded his images made in the city of Bamberg and calculated a recommendation on the rest of his images. If we are not able to get a recommendation we did not take it under consideration when calculating the precision value. For the first three measures of section 4 (SFS, TFS and TIS) we need at least one overlap on one of the $CSDs$ with another user otherwise the similarity between the initiator and all other users is zero and the recommendation will be empty. In the case of the geographic coverage similarity we can always find a similar user as all decisions have at least the world node in common. Using this evaluation method we tested the recommendation on a collected test set of 31 users that made nearly 500 images in Bamberg. Every user made 15 images in Bamberg in average which are scattered in average among eight different spatial choices per user.

As performance metrics you usually find precision (and recall) in various evaluations of recommender systems in conjunction with a cross-validation [14] [11] [12] [17]. Precision is defined as relevant items divided by the number of recommended items. Because always 10 items are recommended, we work with precision at rank 10, or P@10 for short. In addition, an evaluation at higher ranks is not really interesting because of the low average count of eight spatial decisions per user. We consider the $CSDs$ found in the test set to constitute the relevant items as we suppose that a user is only uploading an item if it is relevant to him. Precision is therefore the percentage of the recommended items that are found in the selection of the user. Recall measures the percentage of discovered relevant items against the count of all relevant items. We take a users own feedback in the hidden test set to evaluate the precision and recall of his recommendation.

The precision and recall of each top-10 recommendation based on the different user similarity measurement can be seen in figures 7a and 7b. The

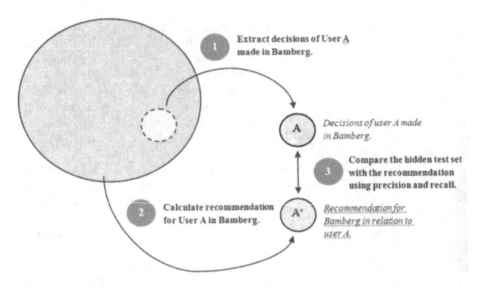

Fig. 6. Steps taken for the cross validation of a single user A

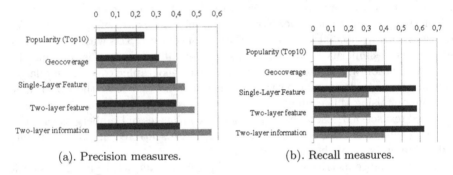

(a). Precision measures.   (b). Recall measures.

Fig. 7. Precision and recall. Black is the precision/recall at rank 10, gray is recommendations without objects of a rank lower than 10

recommendation approach performs significantly better than the baseline. Reading the figure we see that each of the proposed similarity measurements is able to improve the precision (black bars in in reffig:image107) from the initial precision of 0.24 up to 0.41. The recall value (black bars in in 7b) enhanced from a value of 0.35 up to 0.62, which indicates a real improvement in combination with the rising precision. The user similarity measurement that evaluated the best results was the two-layer information similarity, which takes the full potential of the generated semantic model into account. The second precision value (gray bars in figure 7a and 7b) describes the precision of the recommendation after excluding every object that was already found in the top 10 from the list of recommended objects. This measure gives a hind how good our approach is in recommending objects that are not seen by a simple popularity ranking, respectively how good the results are in the long tail. The best values for precision as well as recall were achieved by the two-layer information similarity, respectively 71% and 75% better than the baseline.

6   Conclusion and Outlook

We identified geographic metadata as a possible user feedback for a geographic recommender system that is able to suggest geographic objects. Based on the feedback we added explicit semantic to a partonomy. We proposed four different user similarity measures based on the spatial choices these users made in different geographic contexts. The evaluation of these similarity measures to support recommendation for georeferenced images from panoramio, showed that the described two-layer information similarity (TIS) provides the best personalization results. In conclusion, we may say that notions of spatial similarity, useful for improving geographic recommending, should take into account data about the frequency of spatial choices mapped to a partonomy. Our approach shows that data from the semantic Web can be combined with data from the social Web to support a recommendation system.

Because of our success with one source of implicit semantics, we believe that there are other as yet undiscovered sources. One promising direction could be the use of the temporal context of spatial choices, such as their order, duration or temporal frequency. Additionally, we also intent to investigate how recommenders could profit from explicit semantics attached to the objects. This could help in better separate the objects into categorized $CSDs$ or to express the semantic behind user-based selections of images.

Recommender techniques offer a variety of different approaches that can be used in conjunction with a spatial partonomy. Item-based collaborative filtering can be used to exploit items that show a high relevance to another. We are able to identify nodes in the partonomy that users most likely associate together. Item similarity allows recommendations in a geographic context like suggesting an additional city in Italy when the user has already visited a few other cities in Italy. Another situation would be the recommendation of places in the immediate environment. Combining recommendation results from various approaches could lead to a hybrid geographic recommendation that answers more advanced queries in the future.

Acknowledgments

The authors gratefully acknowledge support by the European Commission who funded parts of this research within the Tripod project under contract number IST-FP6-045335. We also wish to thank Neil Crossley for his fruitful discussions about geographic recommendations.

References

1. Goodchild, M.: Citizens as sensors: the world of volunteered geography. GeoJournal 69(4), 211–221 (2007)
2. Scharl, A., Tochtermann, K., Jain, L., Wu, X.: The Geospatial Web: How Geobrowsers, Social Software and the Web 2.0 are Shaping the Network Society. Springer-11645 /Dig. Serial. Springer, London (2007)
3. Schlieder, C.: Modeling collaborative semantics with a geographic recommender. In: Hainaut, J.-L., Rundensteiner, E.A., Kirchberg, M., Bertolotto, M., Brochhausen, M., Chen, Y.-P.P., Cherfi, S.S.-S., Doerr, M., Han, H., Hartmann, S., Parsons, J., Poels, G., Rolland, C., Trujillo, J., Yu, E., Zimányie, E. (eds.) ER Workshops 2007. LNCS, vol. 4802, pp. 338–347. Springer, Heidelberg (2007)
4. Schlieder, C., Matyas, C.: Photographing a city: An analysis of place concepts based on spatial choices. Spatial Cognition & Computation 9(3), 212–228 (2009)
5. Resnick, P., Iacovou, N., Suchak, M., Bergstrom, P., Riedl, J.: Grouplens: an open architecture for collaborative filtering of netnews. In: CSCW 1994: Proceedings of the 1994 ACM conference on Computer supported cooperative work, pp. 175–186. ACM, New York (1994)
6. Girardin, F., Fiore, F.D., Blat, J., Ratti, C.: Understanding of tourist dynamics from explicitly disclosed location information. In: The 4th International Symposium on LBS & TeleCartography (2007)
7. Ahern, S., Naaman, M., Nair, R., Yang, J.H.I.: World explorer: visualizing aggregate data from unstructured text in geo-referenced collections. In: JCDL 2007: Proceedings of the 7th ACM/IEEE-CS joint conference on Digital libraries, pp. 1–10. ACM, New York (2007)
8. Rattenbury, T., Naaman, M.: Methods for extracting place semantics from flickr tags. ACM Trans. Web 3(1), 1–30 (2009)
9. Snavely, N., Seitz, S.M., Szeliski, R.: Modeling the world from internet photo collections. Int. J. Comput. Vision 80(2), 189–210 (2008)
10. Simon, I., Seitz, S.M.: Scene segmentation using the wisdom of crowds. In: Forsyth, D., Torr, P., Zisserman, A. (eds.) ECCV 2008, Part II. LNCS, vol. 5303, pp. 541–553. Springer, Heidelberg (2008)
11. Burke, R.: Hybrid recommender systems: Survey and experiments. User Modeling and User-Adapted Interaction 12(4), 331–370 (2002)
12. McLaughlin, M.R., Herlocker, J.L.: A collaborative filtering algorithm and evaluation metric that accurately model the user experience. In: SIGIR 2004: Proceedings of the 27th annual international ACM SIGIR conference on Research and development in information retrieval, pp. 329–336. ACM, New York (2004)
13. Zhang, J., Pu, P.: A recursive prediction algorithm for collaborative filtering recommender systems. In: RecSys 2007: Proceedings of the 2007 ACM conference on Recommender systems, pp. 57–64. ACM, New York (2007)
14. Karypis, G.: Evaluation of item-based top-n recommendation algorithms. In: CIKM 2001: Proceedings of the tenth international conference on Information and knowledge management, pp. 247–254. ACM, New York (2001)
15. Park, Y.J., Tuzhilin, A.: The long tail of recommender systems and how to leverage it. In: RecSys 2008: Proceedings of the 2008 ACM conference on Recommender systems, pp. 11–18. ACM, New York (2008)
16. Zhang, M., Hurley, N.: Avoiding monotony: improving the diversity of recommendation lists. In: RecSys 2008: Proceedings of the 2008 ACM conference on Recommender systems, pp. 123–130. ACM, New York (2008)

17. Ziegler, C.N., Lausen, G., Schmidt-Thieme, L.: Taxonomy-driven computation of product recommendations. In: CIKM 2004: Proceedings of the thirteenth ACM international conference on Information and knowledge management, pp. 406–415. ACM, New York (2004)
18. Linden, G., Smith, B., York, J.: Amazon.com recommendations: Item-to-item collaborative filtering. IEEE Internet Computing 7(1), 76–80 (2003)
19. Tversky, A.: Features of similarity. Psychological Review 84, 327–352 (1977)
20. Jones, C.B., Alani, H., Tudhope, D.: Geographical information retrieval with ontologies of place. In: Montello, D.R. (ed.) COSIT 2001. LNCS, vol. 2205, pp. 322–335. Springer, Heidelberg (2001)
21. Rodríguez, M.A., Egenhofer, M.J.: Comparing geospatial entity classes: An asymmetric and context-dependent similarity measure. International Journal of Geographical Information Science 18, 229–256 (2004)
22. Schwering, A.: Approaches to semantic similarity measurement for geo-spatial data: A survey. Transactions in GIS 12(1), 5–29 (2008)
23. Janowicz, K., Raubal, M., Schwering, A., Kuhn, W. (eds.): Special Issue on Semantic Similarity Measurement and Geospatial Applications. Transactions in GIS 12(6) (2008)
24. Resnik, P.: Semantic similarity in a taxonomy: An information-based measure and its application to problems of ambiguity in natural language. Journal of Artificial Intelligence Research 11, 95–130 (1999)
25. Bae-Hee, L., Heung-Nam, K., Jin-Guk, J., Geun-Sik, J.: Location-based service with context data for a restaurant recommendation. In: Bressan, S., Küng, J., Wagner, R. (eds.) DEXA 2006. LNCS, vol. 4080, pp. 430–438. Springer, Heidelberg (2006)
26. Horozov, T., Narasimhan, N., Vasudevan, V.: Using location for personalized poi recommendations in mobile environments. In: SAINT 2006: Proceedings of the International Symposium on Applications on Internet, Washington, DC, USA, pp. 124–129. IEEE Computer Society, Los Alamitos (2006)
27. Tanimoto, T.T.: An Elementary Mathematical Theory of Classification and Prediction (1958)
28. Rosch, E.: Principles of Categorization, pp. 27–48. John Wiley & Sons Inc., Chichester (1978)
29. Guy, M., Tonkin, E.: Folksonomies: Tidying up tags? D-Lib Magazine 12 (2006)
30. Anderson, C.: The Long Tail: Why the Future of Business Is Selling Less of More. Hyperion (2006)
31. Herlocker, J.L., Konstan, J.A., Terveen, L.G., Riedl, J.T.: Evaluating collaborative filtering recommender systems. ACM Trans. Inf. Syst. 22(1), 5–53 (2004)

SPARQL Query Re-writing Using Partonomy Based Transformation Rules*

Prateek Jain[1], Peter Z. Yeh[2], Kunal Verma[2],
Cory A. Henson[1], and Amit P. Sheth[1]

[1] Kno.e.sis, Computer Science Department, Wright State University,
Dayton, OH, USA
{prateek,cory,amit}@knoesis.org
[2] Accenture Technology Labs,
San Jose, CA, USA
{peter.z.yeh,k.verma}@accenture.com

Abstract. Often the information present in a spatial knowledge base is represented at a different level of granularity and abstraction than the query constraints. For querying ontology's containing spatial information, the precise relationships between spatial entities has to be specified in the basic graph pattern of SPARQL query which can result in long and complex queries. We present a novel approach to help users intuitively write SPARQL queries to query spatial data, rather than relying on knowledge of the ontology structure. Our framework re-writes queries, using transformation rules to exploit part-whole relations between geographical entities to address the mismatches between query constraints and knowledge base. Our experiments were performed on completely third party datasets and queries. Evaluations were performed on Geonames dataset using questions from National Geographic Bee serialized into SPARQL and British Administrative Geography Ontology using questions from a popular trivia website. These experiments demonstrate high precision in retrieval of results and ease in writing queries.

Keywords: Geospatial Semantic Web, Spatial Query Processing, SPARQL, Query Re-writing, Partonomy, Transformation Rules, Spatial information retrieval.

1 Introduction

Recently, spatial information has become widely available to consumers through a number of popular sites such as Google Maps, Yahoo Maps and Geonames.org [1]. In the context of the Semantic Web, Geonames has provided RDF [2] encoding of their knowledge base. One issue that makes using the Geonames ontology, or any non-trivial spatial ontology difficult to use, is that users have to completely understand the structure of the ontology before they can write meaningful queries. To illustrate our point, consider the following query from National Geographic Bee [3], "In which

---

* The evaluation components related to this work are available for download at
http://knoesis.wright.edu/students/prateek/geos.htm

K. Janowicz, M. Raubal, and S. Levashkin (Eds.): GeoS 2009, LNCS 5892, pp. 140–158, 2009.
© Springer-Verlag Berlin Heidelberg 2009

country is the city of Pamplona?" This seems to be a straightforward question, and one would assume that the logic for encoding this question into SPARQL [4] query would be to ask – Return a country which contains a city called Pamplona. However, it turns out that such a simple query does not work. This is because Pamplona is a city within a state, within the country of Spain. Therefore the correct logic for encoding the question into query would be – Return a country which contains a state, which contains a county, which contains a city called Pamplona. Unless the user fully understands the structure of the ontology, it is not possible to write such queries.

In this paper, we describe a system called PARQ (Partonomical Relationship Based Query Rewriting System) that will automatically align the gap between the constraints expressed in user's query and the actual structured representation of information in the ontology. We leverage existing work in classification of partonomic relationships[5] to re-write queries.

To study the accuracy of our approach for re-write, we tested it on (1) 120 randomly selected questions from the National Geographic Bee and evaluated them on Geonames ontology (2) 46 randomly selected trivia questions related to British villages and counties from trivia website[22] and evaluated them on British Administrative Geography Ontology[23]. For both the evaluations, users were instructed to read the questions and to write queries in SPARQL for the questions. PARQ rewrote the queries using partonomical relationships. The results were encouraging, and on an average, for evaluation 1, PARQ was able to re-write and answer 84 of 120 queries posed by users, whereas a SPARQL processing system could answer only 20 such queries. For evaluation 2, PARQ was able to re-write and answer 41 of 46 queries posed by users. For both the evaluations, we also compare the performance of PARQ with another well known system PSPARQL [24] which extends SPARQL with path expressions to allow use of regular expressions with variables in predicate position of SPARQL.

The contributions of this work are the following:

1. This work focuses on rewriting SPARQL Queries, written from a user's perspective without worrying about the underlying representation of information.
2. Our work utilizes partonomic transformation rules to re-write SPARQL queries.
3. PARQ has been completely evaluated on third party data (queries and dataset) and shows that it is able to re-write and answer queries not answered by a SPARQL processing system. We demonstrate PARQ can significantly improve precision without any recall loss.

The rest of the paper is organized as follows: section 2 discusses the background work, section 3 discusses approach followed by evaluation in Section 4. In Section 5, we discuss the related work and finally we conclude with section 6.

2 Background

All spatial entities are fundamentally part of some other spatial entity. Hence, spatial query processing systems often encounter queries such as (1) querying for parts of spatial entities (for example, give me all counties in Ohio) (2) querying for wholes which encompass spatial parts (for example, return a country which contains a city called Pamplona).

By identifying which relationships between spatial entities are partonomic in nature it becomes feasible to identify if queries involving those relationships fail because of part-whole mismatch and it becomes possible to fix the mismatches using transformation rules that leverage the partonomic relationships. In this section, we will provide a brief overview of work related to partonomic relationships.

Our work of query rewriting to remove these mismatches is based upon using well-accepted partonomic relationships to address mismatches between a user's conceptualization of a domain and the actual information structure.

Part/Whole relation, or partonomy, is an important fundamental relationship which manifests itself across all physical entities such as human made objects (Cup-Handle), social groups (Jury-Jurors) and conceptual entities such as time intervals (5th hour of the day). Its frequent occurrence results in manifestation of part-for-whole mismatch and whole-for-part mismatch within many domains especially spatial datasets.

Winston [5] created a categorization of part whole relations which identified and covers part whole relations from a number of domains such as artifacts, geographical entities, food and liquids. We believe it is one of the most comprehensive categorization of partonomic relationships and other works in similar spirit such as [6] analyze his categorization.

This categorization has been created using three relational elements:

1. Functional/Non-Functional (F/NF):- Parts are in a specific spatial/temporal relation with respect to each other and to the whole to which they belong. Example: Belgium is a part of NATO partly because of its specific spatial position.
2. Homeomerous/Non-Homeomerous (H/NH):- Parts are same as each other and to the whole. Example: Slice of pie is same as other slices and the pie itself [5].
3. Separable/Inseparable (S/IN): - Parts are separable/ inseparable from the whole. Example: A card can be separated from the deck to which it belongs.

Table 1 illustrates these six different categories, their description using the relational elements and examples of partonomic relationships covered by them.

Using this classification and relational elements, relations between two entities can be marked as partonomic or non partonomic in nature. Further if they are partonomic, the category to which they belong is identified. Finally, appropriate transformation rules can be defined for each category to fix these mismatches.

For the purpose of this work, we have focused our attention on the last category "Place-Area". Places are not parts of any area because of any functional contribution to the whole, and they are similar to the other places in the area as well. Also places cannot be separated from the area to which they belong. Hence, this classification can allow appropriate ontological relationships to be mapped to Place-Area category such as those found in Geonames.

Table 1. Six type of partonomic relation with relational elements

Category	Description	Example
Component-Integral Object	Parts are functional, non-homeomerous and separable from the whole.	Handle-Cup
Member-Collection	Parts are non functional, non homeomerous and separable from the whole.	Tree-Forest
Portion-Mass	Parts are non functional, homeomerous and separable from the whole.	Slice-Pie
Stuff-Object	Parts are non functional, non-homeomerous and not separable from the whole.	Gin-Martini
Feature-Activity	Parts are functional, non-homeomerous and not separable from the whole.	Paying-Shopping
Place-Area	Parts are non functional, homeomerous and not separable from the whole.	Everglades-Florida

3 Approach

At the highest level of abstraction, PARQ takes in a SPARQL query and transforms it with the help of transformation rules. This section provides the details of our system. We describe the various modules of the system, the technologies used for building the system, the transformation rules utilized for transformation of the SPARQL queries and the motivation behind them. Finally we describe the underlying algorithm that explains how the transformation rules are utilized by PARQ for re-writing queries.

3.1 System Architecture

PARQ consists of following three major modules: 1) Mapping Repository 2) Transformation Rule generator and 3) Query Re-writer. Figure 1 illustrates the overall architecture of this system.

Mapping Repository. This module stores mappings of ontological properties to Winston's categories. These mappings are utilized by the Transformation Rule Generator to generate domain specific rules, which are consumed by the Query Re-writer. This is the only module in our system which requires user interaction (other than for query submission). In other words, the user has to specify these mappings.

Each mapping is encoded as a rule in Jena's rule engine format where the antecedent is a triple specifying an ontological property to be mapped and the consequent is a triple specifying the Winston category that the property is mapped to. For example, the following mapping:

[parentFeature: (?a geo:parentFeature ?b)=>(?a place_part_of ?b)]

maps "parentFeature" – a property from the Geonames ontology – to "place_part_of" – Winston's category of Place-Area.

Transformation Rule Generator. This module automatically generates domain specific transformation rules using the mapping repository and pre-defined meta-level transformation rules based on Winston's categories of part-whole relations, which we will explain later. For example, given the following meta-level transformation rule:

[transitivity_placePartOf: (?a   place_part_of   ?b)(?b   place_part_of   ?c)=>(?a place_part_of ?c)]

This module will utilize the parentFeature mapping defined above to generate the following domain specific transformation rule.

[transitivity_parentFeature: (?a geo:parentFeature ?b)(?b geo:parentFeature ?c)=>(?a geo:parentFeature ?c)]

The resulting rule is used by the Query Re-writer to re-write the graph pattern of SPARQL queries in the event of a partonomic mismatch.

This design enables PARQ to be easily used with a wide-range of ontologies. The knowledge engineer only needs to specify the mappings between properties of these ontologies and Winston's categories, which requires less effort than generating the domain-specific transformation rules themselves. This design also allows the transformation rules to be extended in an ontology agnostic manner.

We implemented this module using Jena's [7] rule engine API. Like the mappings, the meta-level transformation rules and the generated rules are encoded in the format accepted by Jena rule engine API. The rule engine allows reading, parsing and processing of rules along with the creation and serialization of new rules.

Query Re-writer. This module re-writes a SPARQL query in case of a partonomic mismatch between the query and the knowledge base to which the query is posed. This module is implemented using Jena and ARQ API [8]. Jena and ARQ provide functionality to convert a query into algebraic representation and vice versa. The triples specified in the query are identified. If they map to partonomic relation using the mapping repository and using Jena's Rule Engine API, the domain specific transformation rule, appropriate transformation is performed on the triples. These transformations are then utilized to re-write the triples exhibiting the mismatch using the features provided by ARQ API.

We believe including transitivity as a part of the reasoner can result in significant overhead for large datasets such as geonames where transitivity applies to almost all the entities. By including it as a part of query rewriting method (1) it allows the mismatches to be resolved on an "on demand" basis (2) it makes it easy to plug in support for resolving other kinds of mismatches.

Original Query

SELECT ?schoolname
{?school geo:parentFeature ?state.   ?state geo:featureCode A.ADM1.
 ?school geo:parentFeature S.SCH. ?school geo:name ?schoolname.
 ?state geo:name "Ohio".}

PARQ System

Mapping Repository

Mappings
1.(a geo:parentFeature b)-> (a place_part_of b)
2.(a lubm:subOrganizationOf b)->(a component_part_of b).
3.(a wine:consistsOf b)->(a stuff_part_of b)
4. .....

Transformation Rules Generator

Meta Level Rules
1. (a place part b),(b place part c)=>(a place part c)
2. (a component part b)(b component part c)=>(a component part c)
3. (a stuff part b)(b stuff part c)=>(a stuff part c)
4...................

Query Re-writer

Domain Specific Rules
1. (a geo:parentFeature b),(b geo:parentFeature c)=>(a geo:parentFeature c)
2.(a lubm:subOrganizationOf b)(b lubm:subOrganizationOf c)=>(a
lubm:subOrganizationOf c)
3. (a wine:consistsOf b),(b wine:consistsOf c)=>(a wine:consistsOf c)
4. ....

Re-written Query

SELECT ?schoolname
{?school geo:parentFeature ?county; geo:featureCode S.SCH;
geo:name ?schoolname. ?county geo:parentFeature ?state.
?state geo:featureCode A.ADM1; geo:name "Ohio".}

Fig. 1. PARQ System Architecture. The relevant rules and mappings for queries shown are
highlighted in bold.

3.2 Meta-level Transformation Rules

Meta-level transformation rules are used to generate domain-specific rules that are used to resolve mismatches resulting from differences in encoding between the granularity of query constraints and the knowledge base by transforming the encoding of the constraints in the query to match the knowledge base.

These meta-level rules are defined at the level of Winston's categories, and a rule defined for a particular category applies to only the partonomic relations covered by that category. For example, rules defined for Component-Object category will cover only relations between machines and their parts, organization and their members, etc.

We used the following methodology to define the meta-level rules used by our system. First, we leveraged previous work by Varzi[9] and Winston, who both showed the semantics of transitivity holds true as long as it is applied across the same category of partonomic relation. From this result, we defined the meta-level transitive transformation rules shown in Table 2, that correspond to Winston's six part-whole categories.

Table 2. Transitivity for Winston's categories

ID	Antecedent1	Antecedent2	Consequent
1	a component part of b	b component part of c	A component part of c
2	a member part of b	b member part of c	A member part of c
3	a portion part of b	b portion part of c	A portion part of c
4	a stuff part of b	b stuff part of c	A stuff part of c
5	a feature part of b	b feature part of c	A feature part of c
6	a place part of b	b place part of c	A place part of c

Next, we investigated the interaction between Winston's categories by examining all possible combinations of these categories for additional transformation rules. This investigation, however, resulted in only frivolous rules, which were not useful for resolving mismatches. For example, the following transformation rule resulted from composing the Feature-Activity category with the Place-Area category.

(a place_part_of b) (b feature_part_of c) => (a feature_part_of c)

However given the following query and triples in an ontology (given in English for brevity),

QUERY: "What state was attacked in WW-II?"

TRIPLE 1 : Florida is a place part of USA (Place-Area).
TRIPLE 2: USA was attacked in WW-II (Feature-Activity)

The rule incorrectly transformed this query to match the ontology, that resulted in an incorrect answer being returned (i.e. Florida).

The reason for these frivolous rules is because Winston's categories are mutually exclusive as they are defined using relational elements. Hence, our meta-level transformations consist of only transitive rules. Despite this small number of rules, we found – through our evaluation – that transitivity by itself provide significant leverage in resolving part-whole mismatches.

3.3 Algorithm

The algorithm used in applying transitivity for resolving mismatches is as follows

SPR= Set of Partonomic Relation
If the query is not well formed
    return
else
    Convert the query Q into its algebric representation (AR).
    Identify the graph pattern(GP) and query variables(QV).
    For every triple t ∈ GP
        if t.property ∈ SPR
            If t.subject is a variable
                Identify other triples with t.subject and use them to unify t.subject
                Insert unified values in s.List
            else
                Insert t.subject in s.List
            If t.object is a variable
                Identify other triples with t.object and use them to unify t.object
                Insert unified values in o.List
            else
                Insert t.object in o.List
        for each s ∈ s.List
        for each o ∈ o.List
        path= Find path between s and o using the transformation rule.
        If (path! =null)
            Replace the resources in the path such that,
                path.source = t.subject.
                path.destination = t.object
The intermediate nodes are replaced such that the object and subject of contiguous triples have the variable names.
Replace the triple in the graph pattern with the path containing the variables.

Return the query Q' to the user

Explanation

Let us explain the algorithm using a query "In which county can you find the village of Crook that is full of lakes?" If the SPARQL Query submitted by user for this question is

```
SELECT  ?countyName
WHERE
  { ?village  ord:hasVernacularName  "Crook" .
    ?county  rdf:type              ord:County ;
              ord:hasVernacularName  ?countyName ;
              ord:spatiallyContains  ?village .

  }
```

Step 1: The system compiles the query to verify if it is well formed. Since, in this case it is a well written query, the system moves on to Step 2.

Step 2: The query is converted into its algebraic representation, and the system iterates through its list of triples to identify triples containing partonomic relationship using the mapping file provided by the user. In this case the last triple

$$t=?county\ ord:spatiallyContains\ ?village$$

contains "spatiallyContains" property which indicates that the object is part of the subject. Hence, this triple is identified as a triple for re-writing.

Step 3: The other triples which contain the variables mentioned in "t",such as:

    ?village ord:hasVernacularName "Crook".,
    ?county rdf:type   ord:County.
    ?county ord:hasVernacularName ?countyName.

are utilized for unifying the values of variables of t (i.e. ?village and ?county). Using these ?village ={ osr7000000000013015 } which is the resource for "Crook" in Administrative Geography Ontology and ?county={set of resources belonging to counties} is computed.

Step 4: The set of unified values from Step3 is then utilized to compute a path by executing transformation rule of transitivity involving the property "tangentiallySpatiallyContains", "completelySpatiallyContains"

?place ={osr7000000000013015}   ?county={List of counties}.This results in the following path being returned:

1. osr7000000000013244 tangentiallySpatiallyContains osr7000000000012934
2. osr7000000000012934 completelySpatiallyContains osr7000000000013015

Step 5: In the path, the source and destination are replaced as mentioned in the original query, and the intermediate node is consistently replaced by a variable.

1. ?county ord:tangentiallySpatiallyContains ?var
2. ?var ord:completelySpatiallyContains ?village.

Step 6: In the original query the last triple is replaced by these two triples resulting in the following query

```
SELECT  ?countyName
WHERE
{
    ?village  ord:hasVernacularName  "Crook" .
    ?county  rdf:type  ord:County ;
             ord:hasVernacularName  ?countyName ;
             ord:tangentiallySpatiallyContains  ?var .
        ?var    ord:completelySpatiallyContains  ?village .
}
```

There can be certain cases where a number of paths are computed between two end points because of transitivity. This will result in generation of multiple re-written queries. We try to rank these generated queries using the following parameters: (1) Re-written queries generating results are given higher ranking than ones which do not (2) If both queries generate results, in those scenarios queries requiring minimum amount of re-writing are given a higher ranking.

4 Evaluation

We present two evaluations to assess the performance of our approach on resolving partonomic mismatches between SPARQL queries written by users and the ontology's to which these queries are posed. We perform these evaluations using: (1) Questions from National Geographic Bee on Geonames Ontology (2) Questions from a popular trivia website which hosts quiz related to "British Villages and Counties" on British Administrative Geography Ontology.

4.1 Evaluation Objective and Setup

Our objective is to determine whether our approach enables users to successfully pose queries about partonomic information to ontology where the users are not familiar with its structure and organization. This lack of familiarity will result in many mismatches that need to be resolved in order to achieve good performance.

To evaluate our objective, we chose Geonames [1] and British Ordinance Survey-Administrative Geography Ontology [23] as our ontology's because: (1) they are one of the richest sources of partonomic information available to the semantic web community. (2) they are rich in spatial information. Geonames has over 8 million place names – such as countries, monument, cities, etc. – which are related to each other via partonomic relationships corresponding to Winston's category of Place-Area. For example, cities are parts of provinces and provinces are parts of countries. Table 3 shows some key relationships found in Geonames.

Table 3. Geonames important properties

Property	Description
http://www.geonames.org/ontology#name	Name of the place
http://www.geonames.org/ontology#featureCode	Identifies if the place is a country, city, capital etc.
http://www.geonames.org/ontology#parentFeature	Identifies that the place identified by domain is located within the place identified by the range

Similarly, Administrative Geography Ontology provides data related to location of villages, counties and cities of the United Kingdom which again map to Winston's place-area relation. Table 4 shows the description of key administrative geography ontology properties. Namespace has been omitted for brevity.

Table 4. Administrative Geography important properties

Property	Description
spatiallyContains	The interior and boundary of one region is completely contained in the interior of the other region, or the interior of one region is completely contained in the interior or the boundary of the other region and their boundaries intersect.
tangentiallySpatiallyContains	The interior of one region is completely contained in the interior or the boundary of the other region and their boundaries intersect. It is a sub-property of spatiallyContains.
completelySpatiallyContains	The interior and boundary of one region is completely contained in the interior of the other region. It is a sub-property of spatiallyContains.

For evaluating our approach on Geonames ontology, we constructed a corpus of queries for evaluation by randomly selecting 120 questions from previous editions of National Geographic Bee[3], an annual competition organized by the National Geographic Society which tests students from across the world on their knowledge of world geography. For British Administrative Geography ontology, we selected 46 questions from a popular trivia website [22] that hosts a number of quizzes related to British geography. We chose these questions for evaluation because:

- These questions are publicly available, so others can replicate our evaluation.
- Each question has a well-defined answer, which avoids ambiguity when grading the performance of our approach.
- These questions are of places and their partonomic relationship to each other. Hence, there is significant overlap with Geonames and Administrative Geography Ontology.

Examples of such questions include:

- The Gobi Desert is the main physical feature in the southern half of a country also known as the homeland of Genghis Khan. Name this country.
- In which English county, also known as "The Jurassic Coast" because of the many fossils to be found there, will you find the village of Beer Hackett?

Once the questions were selected, we employed 4 human respondents (computer science students at a local university) to encode the corresponding SPARQL query for each question. These respondents are familiar with SPARQL (familiarity ranged from intermediate to advanced) but are not familiar with Geonames or Administrative Geography Ontology. These two conditions meet our evaluation objective.

For the National Geographic Bee questions, each subject was given all 120 questions along with a description of the properties in the Geonames ontology. Each subject was then instructed to encode the SPARQL query for each question using these properties and classes.

For the trivia questions, we employed only one human respondent to encode the corresponding SPARQL query because of limitations in time and resources. This respondent was given all 46 questions along with a description of the properties in the administrative geography ontology.

These instructions, original queries, responses and our source code is available for download at http://knoesis.wright.edu/students/prateek/geos.htm

4.2 Geonames Results and Discussion

We compared our approach to PSPARQL and SPARQL. PSPARQL [24] extends SPARQL with path expressions to allow use of regular expressions with variables in predicate position of SPARQL. The regular expression patterns allowed in PSPARQL grammar can be constructed over the set of uris, blank nodes and variables. For example, the following query when posed to PSPARQL returns all cities connected to the capital of France by a plane or train.

```
Select ?City2
WHERE
{  ?City1 ex:capital ex:France .
   ?City1 (ex:plane | ex:train) ?City2 . }
```

We posed queries encoded by human respondents (see previous subsection) to SPARQL and PARQ. We graded the performance of each approach using the metrics of precision (i.e. the number of correct answers over the total number of answers given by an approach) and recall (i.e. the number of correct answers over the total number of answers for the queries). We said an approach correctly answered a query if its answer was the same as the answer provided by the National Geographic Bee.

Table 5 shows the result of this evaluation for PARQ and SPARQL. PARQ on an average correctly re-writes 84 queries of the 120 posed by users performing significantly better than SPARQL processing system across all respondents ($p < 0.01$ for the $X2$ test in each case). The low performance (61 queries by using PARQ and 19 by

Table 5. Comparison Re-written queries Vs original SPARQL queries

	System	# of queries answered	Precision	Recall
Respondent1	PARQ	82	100%	68.3%
	SPARQL	25	100%	20.83%
Respondent2	PARQ	93	100%	77.5%
	SPARQL	26	100%	21.6%
Respondent3	PARQ	61	100%	50.83%
	SPARQL	19	100%	15.83%
Respondent4	PARQ	103	100%	85.83%
	SPARQL	33	100%	27.5%

SPARQL) for respondent 3 can be attributed to this subject having the least familiarity with writing queries in SPARQL and writing improper SPARQL queries. The high performance (103 queries using PARQ and 33 using SPARQL) for respondent 4, can be attributed to this subject having the most experience with SPARQL. For each respondent, the difference of 120 and re-written queries is the number of queries not re-written using PARQ.

For this comparison, we also compared the execution time of PARQ to PSPARQL as shown in Table 6. Because of limitations in time and resources, we were able to employ only one respondent to encode the queries posed to PSPARQL. Hence, we selected Respondent 4 because this respondent has the most experience and familiarity with SPARQL.

Table 6. Comparison PSPARQL and PARQ for Respondent 4

System	Precision	Recall	Execution time/query in seconds
PARQ	100%	86.7%	0.3976
PSPARQL	6.414%	86.7%	37.59

Although PARQ and PSPARQL deliver the same recall (86.7%), we clearly illustrate that PARQ performs much better than PSPARQL in precision ($p<0.01$ for X2 test) because of retrieval of multiple answers by PSPARQL even when the particular resource was present only once in the ontology, thus exhibiting a flaw in the underlying algorithm or implementation. It also illustrates that PSPARQL takes almost 95% more time on, average in answering a query than PARQ ($p<0.05$ for 2-tailed pair-wise t-test).

These results shows that mismatches are common when posing queries to an ontology and that our approach can successfully resolve these mismatches which enabled more queries to be correctly answered.

For example, given the question:

"In which country is Grand Erg Oriental?"

Most of the subjects produced the following query.

```
PREFIX geo:<http://www.geonames.org/ontology#>
SELECT ?countryname
WHERE
    {       ?country    geo:featureCode geo:A.PCLI.
                        geo:name ?countryname.
            ?place      geo:name "Grand Erg Oriental";
                        geo:parentFeature ?country.}
```

This query, however, failed to return any results when posed to Geonames because in Geonames "Grand Erg Oriental" is represented as a part of "Tunis al Janubiyah Wilayat" (a state) which is a part of "Tunisia" (a country). PARQ was able to re-write the original query to align with Geonames (see rewritten query below) which enabled the correct result to be retrieved (i.e. Tunisia).

```
PREFIX geo:<http://www.geonames.org/ontology#>
SELECT ?countryname
WHERE
    {
            ?country geo:featureCode geo:A.PCLI;
                    geo:name ?countryname.
            ?place  geo:name "Grand Erg Oriental".
                    geo:parentFeature ?var.
            ?var    geo:parentFeature ?country.
    }
```

4.3 Administrative Geography Ontology Results and Discussion

For the questions related to British villages and counties, we also compared our approach to PSPARQL. We did not compare our approach to SPARQL because it delivered poor performance in the previous evaluation. Because of time and resource limitations, we were able to employ only one respondent to serialize trivia questions related to British Villages for PARQ and PSPARQL. Again, we selected Respondent 4 for this task because this respondent has the most experience and familiarity with SPARQL, The performance of each approach was graded using precision and recall, and we also compared the execution time of both approaches. We said an approach correctly answered a query if its answer was the same as the answer provided by the trivia website. As illustrated in Table 7 PSPARQL and PARQ perform equally well for recall, but PARQ has a much better precision than PSPARQL ($p<0.01$ for X2 test). It also illustrates PSPARQL on an average is 28 times slower than PARQ ($p<0.05$ for the 2-tailed pair-wise t-test).

Table 7. Comparison PSPARQL and PARQ for Respondent 4

System	Precision	Recall	Execution time/query in seconds
PARQ	100%	89.13%	0.099
PSPARQL	65.079%	89.13%	2.79

These results again illustrate the fact that part-for-whole and whole-for-part mismatches are common in spatial ontology's and PARQ helps resolve these mismatches allowing users to write queries without worrying about the structure of the ontology. As for example for the following trivia question "In which English county, also known as "The Jurassic Coast" because of the many fossils to be found there, will you find the village of Beer Hackett?".

The user poses the following SPARQL query for the question (Namespace omitted for brevity).

```
SELECT  ?countyName
WHERE
{ ?village  ord:hasVernacularName  "Beer Hackett" .
  ?county  rdf:type            ord:County ;
           ord:hasVernacularName  ?countyName ;
           ord:spatiallyContains  ?village .
}
```

The above specified query will not fetch any results because (1) the instance data for Administrative Geography models information using two subproperties of spatiallyContains namely "tangentiallySpatiallyContains" and "completelySpatiallyContains". (2) Villages may or may not be directly part of counties and may contain additional administrative divisions in between.

Unfortunately the difference between "tangentiallySpatiallyContains" and "completelySpatiallyContains" is very subtle and makes it extremely difficult for a naïve user to correctly identify and use the property for querying the ontology, unless the user looks at the instance data and identifies the properties. However, the property "spatiallyContains" is a parent property of both "tangentiallySpatiallyContains" and "completelySpatiallyContains" and is perhaps the most intuitive property of the ontology which captures the semantics of both the properties and can be used by a user for posing queries. So when the above mentioned query is re-written by PARQ according to ontology as following, it retrieves the correct result of "Dorset".

```
SELECT  ?countyName
WHERE
  { ?village  ord:hasVernacularName  "Beer Hackett" .
    ?county  rdf:type  ord:County ;
             ord:hasVernacularName  ?countyName ;
             ord:tangentiallySpatiallyContains ?var .
    ?var     ord:completelySpatiallyContains ?village .
  }
```

4.4 Summary of Results and Limitations

Based on our experiments performed we have demonstrated that PARQ significantly improves precision without any loss in recall and performs significantly faster as well over other systems.

Although our approach significantly improved performance over PSPARQL and SPARQL, there were several queries that it could not answer. Our analysis uncovered the following reasons:

- Several queries (e.g. those about political entities) could not be answered because of insufficient information in Geonames. Example of such queries includes "The Cayman Islands are a territory of which country?"

- Some queries required additional transformations beyond the ones we have identified. These transformations involve relations such as containment and overlap of entities which cannot be defined in terms of Winston's categories. Hence, we need to extend Winston's categories to handle these types of mismatches. Example of such queries includes "Which continent contains the largest number of landlocked countries?"

- Some questions required features, such as aggregate functions, that are not part of the standard SPARQL specification. Our current focus is to provide support for features which are part of standard SPARQL specification. Example of such queries includes "Not including Taiwan, how many provinces comprise China?"

5 Related Work

To the best of our knowledge this is the first work which tries to allow users to formulate SPARQL queries from their perspective without having to worry about the structure of the ontology. However, there are existing works related to RDF Query processing and retrieval of spatial information some of which we think are worth mentioning to highlight their salient features and distinguish our work from them.

The use of Semantic Web technologies for better retrieval of spatial information by incorporating data semantics and exploiting it during the search process was illustrated in [26]. Building upon the vision of [26], for retrieval of spatial information, in our previous work [10] we have defined operators to query spatial, temporal and thematic information from RDF datasets. Our approach for retrieval of spatial information in that work utilizes metric parameters such as geometric co-ordinates, radius, buffer for defining various operators. The operators enhance the standard spatial operators provided by Oracle Spatial and are implemented as supplemental to SPARQL. The reliance on metric parameters compliments our approach here which relies on utilization of named relationships.

Another interesting approach for querying spatial information using SPARQL[11] advocates re-modeling of ontology, than extending SPARQL for retrieval of information. Because of the emphasis on remodeling ontology than transformation of query, this work is obviously along a different dimension than our work. But the work discusses shortcomings of SPARQL for querying spatial data and discusses some interesting query types which a language tailored for spatial querying should be able to handle and hence motivates us in our work. In [17] authors discuss a system for storing spatial and semantic web data efficiently without sacrificing query efficiency which in future can help us in supporting various other kinds of queries.

In [12] we have defined operators for identifying paths in RDF dataset given a source and destination. Using these operators it is possible to express constraints such as the length of the path, specifying a particular node to include in the paths etc. Our current work differs from these works since this work is not on identifying paths. Additionally, our system re-writes SPARQL queries and does not require specification of source and destinations for results to be retrieved. In [25][24] investigate incorporation of regular expressions in the predicate position of SPARQL queries. Though some of these works can be used for answering the queries they suffer from issues of poor precision and slower execution time as demonstrated through our evaluation. Query re-writing has been investigated in other research areas such as databases for yielding better execution plans, data integration and semantic data caching in client-server system [19]. In context of query languages for structured graph data models, [20][21] deal with queries that involve transitive or repetitive patterns of relations in context of databases.

There has been work in spatial query processing system for retrieval of information using partonomic relation such as in [13][18], but not in the context of SPARQL and not utilizing named relationships. These works rely on the use of metric relations such as radius, distance etc. [13] focus on creation of composite or higher order objects via the process of thematic and spatial abstraction.

The work which comes close to our approach is [14]. The work utilizes OWL-DL entailment rules for re-writing SPARQL to retrieve inference results. Unlike our approach where we alter the original graph pattern, the queries are altered by extending graph pattern using UNION construct of SPARQL. In the absence of an accessible implementation, it becomes difficult to compare our approach with the system.

Another work SPARQL-DL[15] incorporates the semantics of SPARQL in their DL reasoner and hence, is along a different dimension than our work.

Some other works on query rewriting are related to Query Optimization [16] , but in our work we are more concerned with retrieval of information from spatial datasets by harnessing partonomic relationships than its optimization.

6   Conclusion and Future Work

We have presented an approach for supporting SPARQL rewriting to allow users to write queries from their perspective without having to worry about the structure of the ontology. Our experiments have been completely performed on third party dataset and queries. Using our experimental results we have proven that our system re-writes these queries using transformation rules such as transitivity effectively and thus helps in resolving the mismatch between query constraints and underlying knowledge base while maintaining a high level of precision of results. Further we have demonstrated that PARQ is significantly faster and can improve precision without any loss to recall.

Our future research ideas include support to handle mismatches that cannot be handled by transitivity alone such as overlap, spatial inclusion. We are investigating support for more SPARQL constructs such as FILTER, OPTIONAL pattern. We are also further testing our approach for its applicability across domains. Limited tests performed show that our approach performs well across other domains of partonomic relations as well. A systematic comparison between resolving mismatches using query re-writing method viz-a-viz a reasoner is part of some of the future goals of this work.

Acknowledgments. This research is funded primarily by NSF Award#IIS-0842129, titled "III-SGER: Spatio-Temporal-Thematic Queries of Semantic Web Data: a Study of Expressivity and Efficiency" and secondarily by NSF ITR Award #071441, "Semantic Discovery: Discovering Complex Relationships in Semantic Web".

References

1. Geonames, http://geonames.org
2. Resource Description Framework, http://www.w3.org/RDF/
3. National Geographic Bee, http://www.nationalgeographic.com/geobee/
4. Prud'hommeaux, E., Seaborne, A.: SPARQL Query Language for RDF (2008), http://www.w3.org/TR/rdf-sparql-query/
5. Winston, M.E., Chaffin, R., Herrmann, D.: A Taxonomy of Part-Whole Relations. Cognitive Science 11, 417–444 (1987)
6. Peter, G., Simone, P.: A conceptual theory of part-whole relations and its applications. Data & Knowledge Engineering 20, 305–3227 (1996)
7. Jena, http://jena.sourceforge.net/
8. ARQ, http://jena.sourceforge.net/ARQ/
9. Varzi, A.C.: A note on the transitivity of parthood. Applied Ontology 1, 141–146 (2006)
10. Perry, M., Sheth, A., Hakimpour, F., Jain, P.: Supporting Complex Thematic, Spatial and Temporal Queries over Semantic Web Data. In: Fonseca, F., Rodríguez, M.A., Levashkin, S. (eds.) GeoS 2007. LNCS, vol. 4853, pp. 228–246. Springer, Heidelberg (2007)
11. Kolas, D.: Supporting Spatial Semantics with SPARQL. Terra Cognita, Karlsruhe (2008)
12. Anyanwu, K., Maduko, A., Sheth, A.P.: SPARQ2L: Towards Support For Subgraph Extraction Queries in RDF Databases. In: 16th World Wide Web Conference (WWW 2007), Banff, Canada (2007)
13. Omair, C., William, A.M.: Utilising Partonomic Information in the Creation of Hierarchical Geographies. In: 10th ICA Workshop on Generalisation and Multiple Representation, Moscow, Russia,
14. Jing, Y., Jeong, D., Baik, D.-K.: SPARQL Graph Pattern Rewriting for OWL-DL Inference Query. In: Fourth International Conference on Networked Computing and Advanced Information Management, NCM 2008 (2008)
15. Sirin, E., Parsia, B.: SPARQL-DL: SPARQL Query for OWL-DL. In: OWLED 2007 Workshop on OWL: Experiences and Directions, Innsbruck, Austria (2007)
16. Hartig, O., Heese, R.: SPARQL Query Graph Model for Query Optimization. In: Franconi, E., Kifer, M., May, W. (eds.) ESWC 2007. LNCS, vol. 4519. Springer, Heidelberg (2007)
17. Kolas, D., Self, T.: Spatially-Augmented Knowledgebase. In: Aberer, K., Choi, K.-S., Noy, N., Allemang, D., Lee, K.-I., Nixon, L.J.B., Golbeck, J., Mika, P., Maynard, D., Mizoguchi, R., Schreiber, G., Cudré-Mauroux, P. (eds.) ASWC 2007 and ISWC 2007. LNCS, vol. 4825, pp. 792–801. Springer, Heidelberg (2007)

18. Vogele, T., Hübner, S., Schuster, G.: BUSTER-an information broker for the semantic web. In: KI, vol. 17, p. 31 (2003)
19. Halevy, A.Y.: Answering queries using views: A survey. VLDB Journal (2001)
20. Cruz, I.F., Mendelzon, A.O., Wood, P.T.: A graphical query language supporting recursion. SIGMOD Record 16, 323–330 (1987)
21. Cruz, I.F., Mendelzon, A.O., Wood, P.T.: G+: Recursive Queries Without Recursion. In: Expert Database Conference (1988)
22. http://www.funtrivia.com/
23. Administrative Geography Ontology, http://www.ordnancesurvey.co.uk/ oswebsite/ontology/AdministrativeGeography/v2.0/ AdministrativeGeography.rdf
24. Alkhateeb, F., Baget, J.-F., Euzenat, J.: Extending SPARQL with regular expression patterns (for querying RDF). Web Semantics 7, 57–73 (2009)
25. Pérez, J., Arenas, M., Gutierrez, C.: nSPARQL: A Navigational Language for RDF. In: Sheth, A.P., Staab, S., Dean, M., Paolucci, M., Maynard, D., Finin, T., Thirunarayan, K. (eds.) ISWC 2008. LNCS, vol. 5318, pp. 66–81. Springer, Heidelberg (2008)
26. Egenhofer, M.J.: Toward the semantic geospatial web. In: 10th ACM international symposium on Advances in geographic information systems. ACM, New York (2002)

iRank: Ranking Geographical Information by Conceptual, Geographic and Topologic Similarity

Felix Mata

Interdisciplinary Professional Unit in Engineering and Advanced Tecnologies
National Polytechnic Institute
Av. IPN No. 2580, Col. Barrio la Laguna Ticomán,
Gustavo A. Madero, CP. 07340, México, D.F.
mmatar@ipn.mx

Abstract. Geographic Information Ranking consists of measuring if a document (answer) is relevant to a spatial query. It is done by comparing characteristics in common between document and query. The most popular approaches compare just one aspect of geographical data (geographic properties, topology, among others). It limits the assessment of document relevance. Nevertheless, it can be improved when key characteristics of geographical objects are considered in the ranking (1) geographical attributes, (2) topological relations, and (3) geographical concepts. In this paper, we outline *iRank* a method that integrates these three aspects to rank a document. Our approach evaluates documents from three sources of information: *GeoOntologies*, dictionaries, and topology files. Relevance is measured according to three stages. In the first stage, the relevance is computed by processing concepts; in second stage relevance is calculated using geographic attributes. In the last stage, the relevance is measured by computing topologic relations. Thus, the main contribution of *iRank* is show that integration of three ranking criteria is better than when they are used in separate way.

Keywords: Geographical Information Ranking, Geographic Information Retrieval, GeoOntology, Spatial Semantics, Topological and Conceptual similarity, Geo-Spatial Relevance.

1 Introduction

Today, many of the mechanisms for calculating the relevance are based on comparing a similarity of a document against a query. For example, the information retrieval (IR) vector model [7] measures document relevance based on word frequency. Web documents relevance is measured using links frequency, while ranking classifies weighted documents by order of importance. In Geographic Information Retrieval, the weighting and ranking mechanisms are based on characteristics of geographic objects. Herein, we select topology, geographical attributes, and spatial semantics because they are broadly used in many GIR and GIS tasks. Therefore, they are aspects ideals for comparing and establish if a document is relevant to a spatial query. However, processing these elements faces two problems: the first one concerns with the diversity of information sources (different encoding, formats, and representations) and the second one – spatial semantics (its processing and storage).

K. Janowicz, M. Raubal, and S. Levashkin (Eds.): GeoS 2009, LNCS 5892, pp. 159–174, 2009.

The first problem can be solved using IR approaches, Geographic Information Systems (GIS), and XML as an interoperability format. The second problem is more complicated because the geographic meaning is expressed in different ways and in different level of detail. For example, semantics of relation like *"near"* needs special treatment according to satisfy user's expectation, then semantics can be captured based on the user's perception, and taking references from other objects or using metrics that assesses the closeness based on parameters of time, distance or perspective. In similar form, other spatial relations like in query "rivers in jalisco" require special treatment because relation "in" could be stored referred to counties or to countries.

Moreover, the same spatial objects are stored based on one of their characteristics according to different sources of information. For example, dictionaries store geographical attributes of airports (coordinates), while the vector files store topological relations of airports (e.g. *adjacency, connect*), and geographical ontologies (*GeoOntologies*) store semantic relations according to certain hierarchy (international-national-local) and how people perceive its geographic scope (*near, next*). Thus, the integration of these three aspects allows assessing a document according to geographical environment and how it is perceived by people, resulting in an enriched ranking. This is the main motivation of present work.

2 Previous Work

The problem of calculating the geographic importance of a document has been treated by processing the similarity between two geographical locations, one associated to the query and another to document [1] and [2]. However, used criteria have resulted insufficient to define the geographical relevance of a document. Basically, this is due to heterogeneity of spatial representations and diversity in interpretations of spatial semantics. For example, in [3] the authors used operators with topological and proximity relations in spatial databases. In [4], Larson proposes a series of geographic operators for queries. A similar approach is adapted in the project SPIRIT [5] having a set of operators for spatial queries, which use different weighting schemes and determine what method should be applied. For example, the concept "near" is measured using the Euclidean distance, while angular difference is used for "north".Other approaches to determining ranking scores, focus on visualizing multiple dimensions of relevance to facilitate human judgment on document relevance [17].

Ontologies have been also applied to measure the relevance of documents of geographic place names, using query expansion approach [9, 11]. The work in [14] proposes a model of a conceptualization of places (ontology) that measures the similarity between a place name and locations. For example, Zaragoza can be referred to Mexico or Spain. Other works combine measures of distances and include semantic relations [10]. However, weight metrics are applied in isolated way or in combination with IR classical approaches, which do not allow evaluate a document adequately, according to geographic domain.

Geographical dictionaries have been used in tasks of recovery and weighting because they contain information that allows attend spatial queries. For example, in [13], three key components of a dictionary are identified: geographic names, location, and properties. Dictionaries have also been used to disambiguate the queries, transforming place names into geographical coordinates. In [15], the topological distance is processed by relations of distance and direction to assess the similarity of spatial scenes. On the other side, in [12], objects are evaluated with respect to its topological similarity.

Summing-up, we did not find works inside of state-of-the art that integrate *GeoOntologies*, dictionaries, and vector data as whole for ranking tasks. This integration, however, would be very useful because these three sources of information stored the same spatial objects with different geographical representations, different encoding, and enriched semantics at different detail level. Then, processing these elements allows establishing appropriate criteria for ranking geographic information and strongly motivates our work. The rest of paper is organized as follows; in section 3 the framework and its modules are described. Section 4 contains the results obtained with our approach. Finally, in section 5 conclusions and future work are outlined.

3 *iRank*: Integral Ranking by Spatial Semantics, Geographic and Topologic Similarity

iRank is an integral approach that ranks geographical documents coming from three information sources: *GeoOntologies*, topological files, and geographic dictionaries. Before, to explain our work, we describe them briefly.

Geontologies were built in a previous work [6], basically they contains concepts linked by spatial relations (near, connect) and semantic relations (hiperonimy and meronimy). Instances present in this GeoOntology are geographical documents obtained from Web and other geographical repositories; the association between concepts and instances was done in manual form.

While Gazetteers contains the descriptions, attributes, properties and names of geographical objects, they were obtained from INEGI[1] and transformed to XML format to this work. TopologyFiles contain the same geographical objects but described by topological relations among them and they were built to *iGIR* system [6].

iRank uses three key characteristics that describe geographical objects, they are: topology, semantics, and geographic properties. *iRank* consists of three stages, in which the geographical relevance between a geographic query and a document is defined. In the first stage, the relevance is computed by finding concepts related to query and document and comparing these concepts to define relevancy; in the second stage geographic attributes associated to document and query are compared to define relevance, and in the last stage, topologic relations related to objects of document and query, are evaluated to define relevancy. The goal is to obtain an integral relevance. We use the following notations to represent a *<query>* (Q_G), "document" (D_G), {concept} (C_G), and [instance] (I_G).

The format of queries used is a triplet *<what, rel, where>*, where geographical object corresponds to *<what>*, while *<where>* is a geographical reference. *<rel>* is a spatial relation between *<what>* and *<where>*. For example, for query *<Cities near*

---

[1] National Institute of Statistics, Geography and Informatics; from Mexico.

Chapala Lake>; *<what>* element corresponds to *<Cities>*, while *<rel>* is the spatial relation <near>, and *<Chapala Lake>* is *<where>*.

The ranking process begins by finding the concepts associated to a query and a retrieved document (using a GIR system [6]). Next, the concepts found are compared to define if the retrieved document is relevant for a query.

This process is done in the following way: *GeoOntology* is explored to find the concepts associated with each element of the query and the retrieved document.

For example, query *<Lakes near Guadalajara>* and the retrieved document "Chapala Lake next Ocotlan" have associated to following concepts {*Water body*} for *<Lakes>*, {*City*} for "Chapala", and {*Municipality*} for "Ocotlan" into *GeoOntology*. Then, to establish relevance of retrieved document, concepts associated to query are compared with concepts associated to document. Comparing is done by using *confusion* metrics [8], which evaluates similarity between a pair of geographical concepts into a *GeoOntology*. Section 3.1 explains in detail this process. The result is named: *conceptual relevance* of a retrieved document to a query. Figure 1 shows modules that compose the framework of *iRank*.

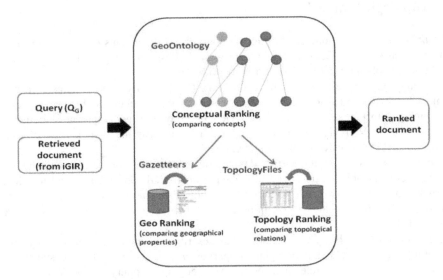

Fig. 1. *iRank* framework

As is shown in Figure 1 the *iRank* framework depicts comparing between query and document retrieved but using three different ranking techniques (conceptual, geographical and topological). These three relevancies are fused (*iRel*) to obtain an integral ranking of geographic documents. *iRel* is calculated using Formula 1:

$$iRel(QG, DG) = \frac{ConRel(Cq, Cd) + GeoRel(Gq, Gd) + TopolyRel(Tq, Td)}{N} \tag{1}$$

where *ConRel* is the value of conceptual relevance between *Cq* and *Cd*. *Cq* is the concept associated to the query and *Cd* is the concept associated to document. For example, "Chapala Lake" vs. "Water body".

GeoRel is the geographical relevance between *Gq* and *Gd*. *Gq* is the concept associated to the query and *Gd* is a geographic attribute of the document. For example, *"Geographic Area"* vs. *"Guadalajara"*.

TopologyRel is the topological relevance. *Tq* is the concept associated to the query and *Td* is a tuple of topologyfile. For example, *"Chapala Lake"* vs *"Lerma River-basin"*.

N is the number of geographic data sources.

iRel is normalized in the range of [0, 1], where one represents the complete relevance and zero corresponds to null relevance. In this way, using the value of integral relevance, the results are weighted to deploy them by ascending or descending order. The rest of section is organized as follows. In section 3.1 conceptual ranking using *GeoOntologies* is explained. Section 3.2 describes how the geographic relevance is measured using gazetteers. In section 3.3, the topological relevance is defined using TopologyFiles. In each section, a *Context Vector* is defined as a mechanism of integration of the rankings. In section 4, the results are shown. Finally, in section 5, our conclusions and future work are discussed.

3.1 Conceptual Ranking

Conceptual ranking is the first stage of *iRank*. This module measures document relevance for a query. Relevancy is determined by comparing concepts associated to present objects in documents. The concepts are found by exploring prebuilt *GeoOntologies*. Where *GeoOntologies* are composed of concepts (rivers, lakes), semantic relations (hyperonimy, meronimy), topological relations (*connect, adjacency*), geographical attributes and instances (an instance is a geographical document). For example, a document about "Lerma River" is an instance of [River] concept). Instances have a property: weight (*Wi*). The value of this property represents the initial relevance of a geographical document. *Wi* is calculated using an approach based on vector model [7], particularly according to keyword frequency. This value is computed for each query; it means that a document has a particular *Wi* related to specific query. A document may describe several geographical objects, even several kinds of geographical objects, then *Wi* should be computed according to these conditions.

The process is as follows: a set of queries were submitted to *Google* and *Yahoo! Answers* and from obtained results (documents), those whose place name match to the label's name of a concept from *GeoOntology* have been selected. This process has been made semi-automatically, using a program developed in Ruby language[2] and criteria of PIIG-Laboratory students and professors.

For example, when *Google* was asked for <*Lakes in Guadalajara*> most of results have been related to "Lerma-Chapala basin." Then, the documents referred to "Lerma-Chapala basin" are considered most relevant (its initial weight is greater) for queries that include <*Lakes in Guadalajara*>. Formula 2 shows how *Wi* is calculated.

$$W_i = \frac{Ft}{Nd} \tag{2}$$

---

[2] Ruby: Language programming. www.ruby-lang.org/es/

where *Wi* is the weight of the concept, *Ft* is occurrence frequency of word associated to the concept, within a document. For its part, *Nd* is the number of considered documents.

The formula normalizes *Wi* into interval [0, 1] so that one represents the maximum value of relevance, while zero is associated to minimum relevance.

Now, we explain how to calculate *conceptual relevance*. For which, we consider the following scenario: a GIS specialist needs to analyze possible flooding in cities near Chapala Lake. Thus, he searches for geographic data using following query: Q_{G1} = <*Cities near Chapala Lake*>. Then, query is processed as follows:

1. - Analyze query to identify each element of the triplet.
2. - Identify the concepts associated to elements of a document and elements of a query.
3. - Extract the context for the document and query.
4. - Process weights (*Wi*) and calculate the *conceptual relevance*.

The first step identifies the elements <*what*>, <*rel*>, and <*where*> of query.

The second step uses the algorithm *OntoExplore* [6] to find in the *GeoOntology* concepts which match to each element of the triplet. For example, for Q_{G1} = <*Cities near Chapala Lake*> *OntoExplore* finds that <Cities> is associated to {Urban_Area}, relation <near> is associated to the concept of {next} and <*Chapala Lake*> to concept {Lake}.

Third step consists of extracting the context of query and document (their neighbor concepts into *GeoOntology*). In this case, for retrieved document D_{G1} = "Ocotlan next Chapala Lake", "Ocotlan" is associated to the concept {Municipality}, while "Chapala Lake" is linked to concept {Lake}, and the relation "next" is associated to {near}. Subsequently, context is extracted and stored in a *Context Vector* (*Vc*). For example, {Water Body} has following neighbor concepts: {Lake} and {River}, which are stored into *Vc*. Figure 2 shows *GeoOntology*, the query, geographic document and *Vc* obtained by *OntoExplore*.

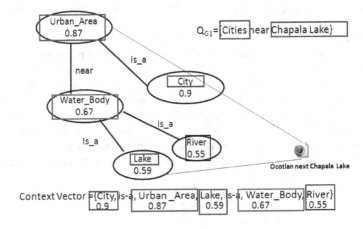

Fig. 2. *Context Vector* obtained by *OntoExplore*

Also, Figure 2 displays concepts, and their weights (inbox).

The fourth and final step is to determine the conceptual relevance between document "Ocotlan next Chapala Lake" and query *<Cities near Chapala Lake>*, for that, the following formula 3 is applied. The value obtained is *conceptual relevance* (*ConRel*).

$$ConRel(ci, cj) = \frac{W_{i1} + W_{i2}}{D} \tag{3}$$

where "ConRel" is the *conceptual relevance* between c_i and c_j (e.g. "*City*" vs. "*State*"). c_i is the concept (from query) and c_j is the concept (from document). For example, $c_i = $ "*City*" for "*Chapala*" and $c_j = $ "*state*" for "*Guadalajara*".

Wi_1 and Wi_2 are the initial weights of the involved concepts. For example, *Chapala Lake* has an initial weight of 0.78 according to the formula 2.

D is the confusion (similarity) between the concepts c_i and c_j. It is calculated by distance between these concepts (at node level). For example, if $c_i = $ Guadalajara and $c_j =$Chapala, then the node that represents "Chapala" is located, and number of nodes required to reach the node "Guadalajara" is defined and it is the value of D.

Therefore, *ConRel* measures similarity between pair of concepts (one concept from document and other concept from query) it applies with each element of query and document. We suppose that document and query have the same number of elements.

Finally *ConRel* of document $D_{GI} = $ "Ocotlan next Chapala Lake" and query *<Cities near Chapala Lake>* is obtained. First ranking stage finishes here. The next task consists of processing context vector to weigh the documents retrieved from two other information sources. This process is explained in section 3.2 (*Geographical Ranking*) and section 3.3 (*Topological Ranking*).

3.2 Geographical Ranking

Geographical Ranking (*GeoRank*) is the second stage of *iRank*. *GeoRank* measures geographic relevance of a retrieved document. Relevance is computed by comparing geographic properties of objects embedded into query and into retrieved document. Geographic properties are searched into gazetteers. The result of comparing is called *geographical relevance (GeoRel)*.

Gazetteers used in this stage were built to previous work [6]. They contain relations, properties and some constraints according to spatial database generated by INEGI from Mexico. They contain information directly related to geographical names that appear on the topographical map in its various scales. Moreover, it offers basic information about the localities represented in the vector component database.

Additionally, these data dictionaries are generated at scales of 1:50,000 and 1:250,000 and the reference vector model. The use of these documents provide the basis for identifying which objects exist in spatial databases generated in Mexico.

These properties and relations are not contained in *GeoOntologies*. The goal is to use other attributes and relations to measure similarity of spatial objects from query and document related to concepts obtained in conceptual ranking.

The process of *GeoRank* is as follows, first, we use context vector (*Vc*) obtained in section 3.1 and identify geographical properties of the objects included in a query. Then, we apply a two steps process:

1. – Form pairs of geographical objects (*GeoObj*). The first object belongs to query and the second one corresponds to document. We consider that document and query contains the same number of objects.
2. – Weights of *GeoObj* are processed.

To explain this process, we consider the query Q_{G1} = <*Cities near Chapala Lake*> and a pair of retrieved documents from dictionaries. The documents are: D_{G2} = "Urban Area shares Railroad" and D_{G3} = "Urban Area shares Water Body". Applying the first step for document D_{G1}, the following pair of objects is formed: "Urban Area" vs. <City>, the relations pair *"share"* vs. <near>, and finally "Railroad" vs. <*Chapala Lake*>. Thus, applying the second step to <*Urban Area*> vs. "City", we have that <*Urban Area*> has a Wi = 0, 87 (see section 3.1); while "City" has a Wi=0, 76. An average between these values is calculated, giving the relevance of <*Urban Area*>. Relation "*shares*" has Wi = 0,7 with respect to the relation <near>. While *"Water Body"* has Wi = 0, 67 and <*Chapala Lake*> has Wi = 0, 87. An average between them is calculated, giving the relevance of *"Water Body"*. The *geographical relevance* (*GeoRel*) is obtained by formula 4.

$$GeoRel(Q_G, D_G) = \frac{W_{c1} + W_{c2} + W_{c3}}{3} \qquad (4)$$

where, *GeoRel* is the geographical confusion between query Q_G and the document (D_G). Wi is initial weight of each element of the triplet <*what, rel, where*>.

Therefore, applying formula 4 to D_{G2}, we obtain: (0.87 +0.7 +0) / 3 = 0.52 While for D_{G3}, we have: (0.87 +0.7 +0.67) / 3 = 0.74. Figure 3 shows the described process for query Q_{G1} and documents D_{G2} y D_{G3} obtained from Gazetteers.

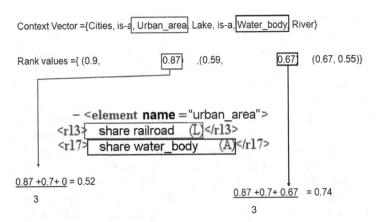

Fig. 3. GeoRanking. Documents retrieved from gazetteers, and *Context Vector* (*Vc*).

Figure 3 shows the context vector and its values of relevance as well as the pair of documents retrieved from dictionaries. Then, *geographical relevance* is calculated by applying formula 4; for example, <*Lake*> vs. "Water Body", <*Lake*> vs. "railroad", and <City> vs. "Urban Area". In that case r13 and r17 are labels of relations number of <*Urban_Area*>.

This is the mechanism of *GeoRank*. This way, second stage of ranking finishes. The next task is to rank retrieved documents from the last information source (*TopologyFiles*). This process is explained in section 3.3 (*Topological Ranking*).

3.3 Topological Ranking

The third stage of *iRank* establishes the topological relevance between document and query, named Topological Ranking (*TopologyRank*). To achieve this, we use *TopologyFiles* [6] (a file format that stores topological relations between two geographic objects). The figure 4 shows the structure of a topologyfile, a spatial query, and a part of context vector obtained by *OntoExplore* algorithm.

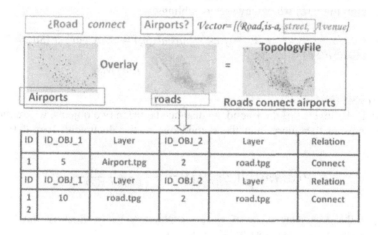

Fig. 4. Structure of *TopologyFile*

In figure 4 the spatial relation *connect* is shown between airports and roads from Mexico. In this case, to assess the relevance, spatial relations are classified into three groups according to what is defined in [1] and [16]. The first group deals with spatial relations of **Inclusion** (if an object A is into an object B), the second one is **Proximity** (how close is the object A to the object B), and the third one are **Siblings** (two concepts are siblings if they have the same father). In the following, we define the rules for assessing these aspects and then rank them with a value of relevance.

3.3.1 Inclusion

Check if *Sd* is within *Sq*, where *Sd* is the geographic scope of the document, while *Sq* is the geographic area of query. Geographic scope of document is geographical area of object involved, it is obtained from TopologyFiles. For example, *Sq* of the query

Q_{GI} is "*Guadalajara*" and *Sd* of the document D_{GI} is "Chapala Lake ". Formula 5 is applied to determine that inclusion between "Guadalajara" and "Chapala Lake" is 2/5.

$$Inclusion\ (Sq, Sd) = \begin{cases} \dfrac{NumDescendants\ (Sd) + \ 1}{NumDescendants\ (Sq) + \ 1} & if\ Sd \sqsubseteq Sq \\ \qquad\qquad 0 & otherwise \end{cases} \tag{5}$$

Formula 5 returns values in the interval [0, 1]. The maximum value is when both elements have the same number of descendants (*Sd* is within *Sq*) and the minimum one when *Sd* has no descendants. **NumDescendants (S) +1** is the number of scopes within S, plus scope itself (that is to say, relations "sub-of-region" in the *GeoOntology*).

3.3.2 Siblings

A binary function checks if *Sq* and *Sd* are siblings in the *GeoOntology*, defined by Formula 6. For example, *"River"* and *"Lake"* have the same father and therefore are siblings. The maximum value (one) of the function is when the elements are siblings and the minimum (zero) when they are not siblings.

$$Siblings(Sq, Sd) = \begin{cases} 1, & if\ \exists\ Sx: parent(Sq) = Sx \land parent\ (Sd) = Sx; \\ 0, & otherwise \end{cases} \tag{6}$$

3.3.3 Proximity

Proximity is the inverse of the Euclidean distance between two objects, where the first object belongs to the query, and the second one to the document. It is defined by formula 7:

$$Proximity(\ Sq, Sd) = \frac{1}{1 + \frac{Distance(Sq, Sd)}{Diagonal\ (Sq)}} \tag{7}$$

where *Sq* is the geographic scope of the geographical reference of query and *Sd* is the geographic scope of the object described by document.

For example, query scope of Q_{GI} is Guadalajara City because Chapala Lake is within Guadalajara City. The allocation of this scope (a numeric value) is obtained manually and automatically using a Java tool and shapefiles in conjunction with criteria established by GIS specialist. In addition, the Euclidean distance is normalized by the diagonal of the MBR (Minimum Bounding Rectangle) defined for geographic area of query (MBR is a rectangle of minimum size that encloses completely the irregular shape of a region).

Now, we proceed to explain how to calculate the topologic relevance with the following example: Considering query Q_{GI} = <*Cities near Chapala Lake*> and a pair of retrieved documents, D_{G4} = "Grijalva River crosses Villahermosa" and D_{G5} = "Ocotlan next Chapala Lake". Note that a priori we know that the document D_{G4} is irrelevant and that the document D_{G5} is relevant for query Q_{GI}. Topologic relevance is calculated using the following four steps: 1. – Check if objects belong to the same class. 2. - Extract the geographic scope of the document and query to assess the

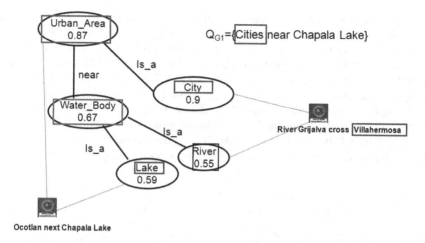

Fig. 5. Identifying concepts associated to the documents and query

proximity, inclusion, and siblings. 3. – Apply an overlay operation between geographic scopes of the document and query. 4. - Topological relevance is calculated by the average of overlay, inclusion, siblings, and proximity.

Then, we have a match between concepts associated to D_{G4} and Q_{G1}, because both of them are linked with {City} concept. Figure 5 shows this process.

The second step is to extract the geographic scope of query and documents. For Q_{G1}, geographic area of <*Chapala Lake*> is extracted. While for document D_{G4} the value of Grijalva River length is extracted. Then, the inclusion, proximity, and siblings for both objects are verified. There is no *inclusion* (*Grijalva River* is not within the *Chapala Lake*) then closeness is zero; *siblings* function is equal to the one, because River and Lake are Water Bodies. *Proximity* is zero. Then, taking the results of these operations, the relevance value is 1/3.

In the third step it is verified if exists overlapping between two objects. If they do not overlap, the topological relevance is zero. In the case of overlapping, the size of overlapped geographical area is defined and this value is considered as its relevance. This operation is displayed in a table where the MBR records of each object are stored. Finally, in the fourth and final step, the results are organized according to the overlapping area in ascending or descending order. The process is the same for the rest of relations associated to proximity, according to the involved relation; previously defined functions are applied to obtain the topological relevance.

4   Experiments and Results

iRank has been tested using documents retrieved by *iGIR* [6] which is a system that retrieves documents based on integral matching using three sources of information (the same as in this paper). We used nine hundred documents, they are selected manually, from which three hundred are topologyfiles [6], another three hundred are elements of geographic dictionaries, and the rest are elements of *GeoOntology*. We considered queries with spatial relations: "near", "connecting", "in", and "within".

The relevance was normalized into interval [0, 1] and we establish five classes for describing the document relevance. The first one named "null relevance" for documents with value = 0. The second one is "small relevance" (values from 0.1 to 0.3), third one range marks "medium relevance" (values from 0.4 to 0.6), fourth one is defined as "somewhat relevant" (values from 0.7 to 0.9). Finally fifth one is "complete relevance" corresponding to the documents weighted with value of one. An example of "complete relevance" is shown using query Q_{G1} = <*Cities near Chapala Lake*> with document D_{G7} = "Guadalajara near Chapala Lake", because Guadalajara is a City near Chapala Lake. To explain results, consider Figure 6 that shows *Chapala Lake*, municipalities, and highways, surrounding it.

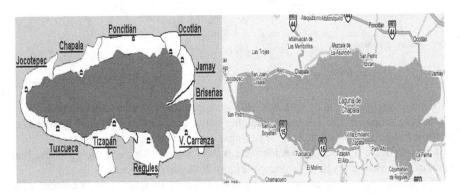

Fig. 6. Counties and highways surrounding *Chapala Lake*

Table 1 shows the results obtained by *iGIR* for query Q_{G1} = <*Cities near Chapala Lake*> and their ranking values according to *GeoRank*, *TopologyRank*, and *Concept Rank (iRank)*.

Table 1. Results for the query: Q_{G1} = <Cities near *Chapala Lake*>

Retrieved Document	GEOR ANK	Rank Position	TOPOLOG Y RANK	Rank Position	CONCEP T RANK	Rank Position	iRank Value	Rank Position
Chapala	0,89	2	0,904	1	0,87	1	0,88	1
Poncitlan	0,96	1	0,891	2	0,42	2	0,75	2
Tizapan	0,83	3	0,837	4	0,26	9	0,64	3
C. Régules	0,79	4	0,810	5	0,25	10	0,62	4
Jocotepec	0,70	6	0,842	3	0,29	3	0,61	5
Tuxcueca	0,75	5	0,673	8	0,29	4	0,57	6
Ocotlan	0,68	8	0,718	6	0,29	5	0,56	7
V.Carranza	0,69	7	0,679	9	0,27	7	0,54	8
Briseñas	0,65	10	0,639	10	0,29	6	0,54	9
Jamay	0,67	9	0,710	7	0,26	8	0.52	10

Table 1 shows rankings corresponding to ten retrieved documents. The last column contains the values generated by *iRank*. For example, the document "Chapala", *GeoRank* places in position 2, while *TopologyRank* and *ConceptRank* place it in the first position. The reason of this difference is that *GeoRank* considers the geographic area of Chapala, which is lower than of Poncitlan municipality. *TopologyRank* considers the roads that connect Poncitlan municipality, and Chapala municipality with Chapala Lake. For its part, *ConceptRank* considers the name of the municipality, which in this case coincides with the name of the lake (synonymy). By integrating these three criteria, *iRank* places it in the first position.

Other example for discussing is the document of "Jocotepec", where each of three weighting measures places it in different position. *GeoRank* places it in the sixth position because its geographic area is the second largest of the ten municipalities. *TopologyRank*, places it thirdly because it has a road that connects with Chapala Lake. *ConceptRank* places it in fourth position according to its semantic relations. By integrating these criteria *iRank* finally places it in fifth position.

Now, we show results in table 2 for query Q_{G2} = {roads connecting *airports*} arranged accord relevance obtained by *iRank*.

As we can see, several documents have opposite relevance values. In other words, according to one ranking criteria a document has high relevance but using other criteria, relevance is medium or low (see seven result of Table 2). To solve these

Table 2. Ranked results for Q_{G2} = {"roads connecting airports"}

Position in iRank	Retrieved document	GEO RANK	Pos	TOPOLOGY RANK	Pos	CONCEPT RANK	Pos	iRank
1	Avenue Boulevard P,Aereo connects B, Juárez airport	0,75	1	0,84	2	0,96	1	0,85
2	Street Carlos León connects B, Juárez airport	0,68	2	0,95	1	0,87	5	0,83
3	Street Eje uno norte connects Terminal 1	0,59	7	0,76	4	0,81	6	0,72
4	Street sonora connects Terminal 2	0,51	8	0,82	3	0,81	2	0,71
5	Street Viaducto Piedad and Río Churubusco connects Terminal 2	0,61	5	0,38	9	0,9	4	0,63
6	Avenue Vial 2, connects Viaducto and Churubusco Terminal 2	0,6	6	0,36	7	0,9	3	0,62
7	Highway connects airport	0,65	3	0,46	8	0,36	9	0,49
8	street connects airport	0,63	4	0,43	5	0,32	10	0,46
9	avenue vial 1 connects Terminal1	0,21	9	0,7	6	0,46	7	0,45
10	Traffic roadways connecting airport	0,14	10	0,12	10	0,46	8	0,24

Pos=Position.

discrepancies *iRank* integrates relevancies for ranking in global way to documents and not only for isolated criteria. It reduces possibility that if a document is relevant according its topology, it will be discarded just only because it was evaluated processing its geographical properties. With our approach level of certainty about relevance of a retrieved document is increased.

Other point to discuss is ranking of documents located in place eight. *GeoRank* and *TopologyRank* evaluated document with a medium relevancy, while *ConceptRank* evaluated with lower relevance. It implies that when semantic relations are processed to establish its relevance, this document will be omitted or ranked in last places, but when its topologic and geographical properties are considered in ranking process (e.g. using *iRank*). Then, document will be located as relevant; it is other example where integration is useful in ranking.

In the following table, we present other results related with query Q_{G3} = <Shopping center within colonies>

Table 3. Ranked and retrieved documents for Q_{G3} = <Shopping center within colonies>

Position in iRank	Retrieved document	GEO RANK	Pos	TOPOLOGY RANK	Pos	CONCEPT RANK	Pos	iRank
1	Shopping center in colonies	0,6	4	0,55	5	0,98	1	0,71
2	aurrera Mall within CTM VII culhuacan	0,45	5	0,65	3	0,94	2	0,68
3	Chedrahui Mall within girasoles I	0,45	7	0,86	9	0,72	6	0,67
4	Diverse installation within urban area	0,74	1	0,51	2	0,68	3	0,64
5	Superama Mall within san pablo tepetlapa	0,73	2	0,93	6	0,26	4	0,64
6	Megacomercial within caracol	0,39	3	0,65	1	0,79	7	0,61
7	Diverse installation within colony	0,7	6	0,36	4	0,61	8	0,55
8	Walmart mall in tepeyac	0,38	8	0,45	8	0,56	10	0,46
9	Oxxo Store within santa ursula	0,37	9	0,75	7	0,18	5	0,43
10	Superama mall in lindavista	0,36	10	0,34	10	0,51	9	0,40

Pos=Position.

In table 3, for query Q_{G3} = {shopping center *within* colonies} result 8 obtains almost the same relevance for three criteria, *iRank* maintains relevance place. In result 5 (document retrieved from Web) according to *GeoRank* document is relevant, but in TopologyRank document is medium relevant. It occurs because *within* relation is not topologically processed in Web documents. Result 3 (which we know is relevant) when is evaluated using only a relevance criteria, it did not obtain a ranking value accord to its relevance. When is evaluated in integral form a ranking value accord to its relevance is assigned.

5   Conclusions and Future Work

We present an integral method for ranking geographic documents obtained from three sources of heterogeneous data: topological files, geographic dictionaries, and *GeoOntologies* (the meaning of geographical space for a group of people) of objects contained in documents and queries.

Similarity between geographic objects is calculated according to their topology, spatial semantics, and geographic properties (integral criteria to rank).

Integration of ranking criteria controls the semantic precision in retrieved documents.

iRank uses the *confusion theory* by taking the advantage of the hierarchical nature of the geographic space, through which one can determine if two objects are similar according to their topology, spatial semantics, and geographic properties. The results show that integrating these aspects the ranking process is improved.

The methodology of *iRank* shows a better ranking compared with other approaches that use some of these three criteria in separated way (see Tables 1, 2 and 3).

The integration of these measures, results in a better ranking because key characteristics of spatial objects are considered in the evaluation of relevancy.

However, additional experiments, using other topological relations, for example relations from model of 9 – intersection, would be very useful in the future work. We plan to enrich *GeoOntologies* with conceptualizations built by GIS communities and Web users. Also, we will design the modules, which process the elements of queries according to place names.

Other future work considers use thematic semantics of the documents. Finally, we need to test the system's performance on larger data collection.

iRank is useful when queries and document retrieved, contain the same spatial objects but they are described by key different characteristics. In addition, when objects retrieved are similar to requested. Then their similarity (relevance) should be measured according to these key characteristics and in integral form.

Acknowledgements

The authors of this paper wish to thank the Centre for Computing Research (CIC), PIIG-Laboratory students and professors, SIP-IPN, National Polytechnic Institute (IPN), and the Mexican National Council for Science and Technology (CONACYT) for their support.

References

1. Egenhofer, M.J., Mark, D.: Naive geography. In: Kuhn, W., Frank, A.U. (eds.) COSIT 1995. LNCS, vol. 988, pp. 1–16. Springer, Heidelberg (1995)
2. Jones, C.B., Alani, H., Tudhope, D.: Geographical information retrieval with ontologies of place. In: Montello, D.R. (ed.) COSIT 2001. LNCS, vol. 2205, p. 322. Springer, Heidelberg (2001)

3. Nedas, K., Egenhofer, M.: Spatial similarity queries with logical operators. In: Hadzilacos, T., Manolopoulos, Y., Roddick, J., Theodoridis, Y. (eds.) SSTD 2003. LNCS, vol. 2750. Springer, Heidelberg (2003)
4. Larson, R.: Geographic information retrieval and spatial browsing. In: Geographic Information Systems and Libraries: Patrons, Maps, and Spatial Information, pp. 81–123 (1995)
5. Vaid, S., Jones, C., Joho, H., Sanderson, M.: Spatio-textual indexing for geographical search on the web. In: Bauzer Medeiros, C., Egenhofer, M.J., Bertino, E. (eds.) SSTD 2005. LNCS, vol. 3633, pp. 218–235. Springer, Heidelberg (2005)
6. Mata, F.: Geographic Information Retrieval by Topological, Geographical, and Conceptual Matching. In: Fonseca, F., Rodríguez, M.A., Levashkin, S. (eds.) GeoS 2007. LNCS, vol. 4853, pp. 98–113. Springer, Heidelberg (2007)
7. Baeza-Yates, R., Ribeiro-Neto, B.: Modern Information Retrieval. ACM Press Series/Addison Wesley (1999)
8. Levachkine, S., Guzman-Arenas, A.: Hierarchy as a new data type for qualitative variables. Expert Systems with Applications: An International Journal 32(3), 899–910 (2007)
9. Jones, C., Abdelmoty, A., Fu, G.: Maintaining ontologies for geographical information retrieval on the web. In: Meersman, R., Tari, Z., Schmidt, D.C. (eds.) CoopIS 2003, DOA 2003, and ODBASE 2003. LNCS, vol. 2888, pp. 934–951. Springer, Heidelberg (2003)
10. Clementini, E., di Felice, P., van Oosterom, P.: A Small Set of Formal Topological Relations Suitable for End-User Interaction. In: Abel, D.J., Ooi, B.-C. (eds.) SSD 1993. LNCS, vol. 692, pp. 277–295. Springer, Heidelberg (1993)
11. Fu, G., Jones, C.B., Abdelmoty, A.I.: Ontology-based spatial query expansion in Information Retrieval. In: Meersman, R., Tari, Z. (eds.) OTM 2005. LNCS, vol. 3761, pp. 1466–1482. Springer, Heidelberg (2005)
12. Belussi, A., Catania, B., Modesta, P.: Towards Topological Consistency and Similarity of Multiresolution Geographical Maps. In: GIS 2005, Bremen, Germany (2005)
13. Hill, L.: Core elements of digital gazetteers: Placenames, categories and footprints. In: Borbinha, J.L., Baker, T. (eds.) ECDL 2000. LNCS, vol. 1923, p. 280. Springer, Heidelberg (2000)
14. Jones, C.B., Harith, A., Tudhope, D.: Geographic Information Retrieval with ontologies of place. In: Montello, D.R. (ed.) COSIT 2001. LNCS, vol. 2205, p. 322. Springer, Heidelberg (2001)
15. Burns, H., Egenhofer, M.: Similarity of Spatial Scenes. In: Proc. 7th Int. Symp. on Spatial Data Handling, pp. 31–42 (1996)
16. Andrade, L., Silva, M.: Relevance Ranking for Geographic IR. In: Workshop on Geographic Information Retrieval, SIGIR, USA (2006)
17. Hobona, G., James, P., Fairbairn, D.: Multidimensional visualisation of degrees of relevance of geographic data. International Journal of Geographic Information Science 20(5), 469–490 (2006)

Towards an Ontology for Reef Islands

Stephanie Duce

Escola Superior de Tecnologia i Cliencies Experimentals, Department of Llanguatges i
Sistems Informatics, University Juame I,
12071 Castellón de la Plana, Spain
Institute for Geoinformatics, and University of Muenster, Germany
stephanie.duce@gmail.com

Abstract. Reef islands are complex, dynamic and vulnerable environments with
a diverse range of stake holders. Communication and data sharing between
these different groups of stake holders is often difficult. An ontology for the
reef island domain would improve the understanding of reef island geomorphology and improve communication between stake holders as well as forming
a platform from which to move towards interoperability and the application of
Information Technology to forecast and monitor these environments. This paper
develops a small, prototypical reef island domain ontology, based on informal,
natural language relations, aligned to the DOLCE upper-level ontology, for 20
fundamental terms within the domain. A subset of these terms and their relations are discussed in detail. This approach reveals and discusses challenges
which must be overcome in the creation of a reef island domain ontology and
which could be relevant to other ontologies in dynamic geospatial domains.

Keywords: Reef Island, Domain Ontology, Informal Ontology, Conceptualizing Dynamic Environments, DOLCE.

1 Introduction

Reef islands are dynamic landforms composed almost entirely of unconsolidated sand
occurring on top of reef flats [1]. They are the combined expression of complex and
inextricably linked geomorphic, chemical and biological processes which occur on
reef flats [2]. Organisms, such as coral, which compose the reef flat produce sediment, which is transported by waves and currents across the reef flat and deposited at
a node of wave refraction on the reef flat [3] (Refer to Figure 1). As the sediment
deposit grows it becomes more stable and may become vegetated, thus, creating a reef
island.

Worldwide, reef islands are home to thousands of people. However, they are small,
low-lying and vulnerable to natural and human-induced environmental changes [4].
Their unique characteristics make the research and management of reef islands important from ecological, social and economic perspectives. However, facilitating effective dialogue, co-operation and interoperability between the many stake holder
groups, including researchers from different backgrounds, managers and local communities is very difficult. The development of an ontology for the reef island domain
could help to overcome some of these difficulties.

K. Janowicz, M. Raubal, and S. Levashkin (Eds.): GeoS 2009, LNCS 5892, pp. 175–187, 2009.
© Springer-Verlag Berlin Heidelberg 2009

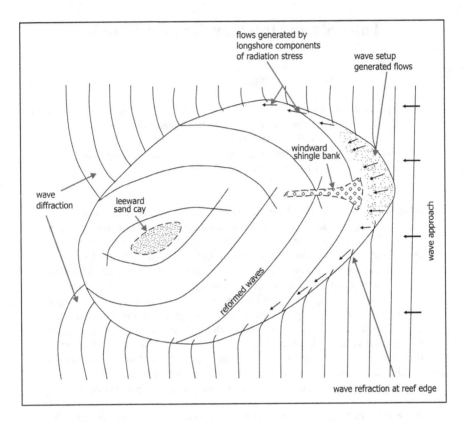

Fig. 1. Sediment deposition on a reef flat and reef island (sand cay) formation at a nodal point of wave refraction (source: 11)

Ontology is broad and diverse with roots in philosophy and now in computer science. In the context of this project, an ontology can be best defined as "a shared vocabulary plus a specification (characterization) of its intended meaning" [5]. Ontologies help to structure knowledge and improve our understanding of concepts of the world by clearly stating how entities relate to one another [6]. When these relationships are formalized (for example in a logical, computer language such as OWL), computers are also able to "understand" and reason about entities and phenomena. By defining entities and their relations, ontologies overcome problems of semantic heterogeneity which are described below.

In a linguistic sense, semantics is the relationship between words and the real world things that the words refer to i.e. the meaning of words or terms. From a Geographic Information System (GIS) perspective, semantics defines the relationships between computer representations and the real world entities which they correspond to in a certain context [7].

Semantic heterogeneity occurs when a word used to refer to something has multiple meanings or can be interpreted differently by people from different domains or backgrounds. For example, what is called a "beach" could differ from a tourist's perspective, who wants a nice place to play in the sand by the water; to the perspective of a geomorphologist, who recognizes the beach as the "active zone of sediment transport" which may continue well below the surface of the water; to that of a biologist, who is interested in the "beach" as a sandy site, above high tide, for turtles to lay eggs.

Naming heterogeneity occurs when the same feature is named using different words. For example, a reef platform may also be called a carbonate platform, reef flat or simply a reef. This problem of semantic heterogeneity was identified by Bishr [7] as one of the biggest barriers to data sharing and interoperability. It also presents a substantial barrier to communication between people particularly from different disciplines of study or from different cultural backgrounds.

A domain ontology specifies the meanings of terms and relationships between entities within a certain field of study. Many disciplines develop standardized ontologies and structured vocabularies which domain experts can use to share and annotate information in their fields [8].

Ontologies are of use for numerous purposes in geographic domains. They can be employed to overcome data management problems by providing a common reasoning framework (e.g. 9). Thus, they also facilitate knowledge sharing and reuse. Chandrasekaran *et al.* [10] state that ontologies need not represent only facts about a given domain but may also represent beliefs, goals, hypotheses and predictions.

1.1 Background and Importance

When communicating coastal issues and their possible solutions to communities it is imperative that there is clear and unambiguous understanding between local community members, coastal scientists and managers. In addition, local community members themselves are often a rich source of information about the dynamics of their area within a historical context and with respect to the daily dynamics. To extract information from local communities and to make the communications of "experts" understandable to locals, a common vocabulary is needed. The meaning of this vocabulary can be specified using an ontology. In addition to improving inter-personal communication between stake-holder groups an ontology can also formalize the semantics of data and metadata structures allowing interoperability and data sharing between organizations and from different sources.

The boundaries of many coastal phenomena and features are indistinct and, given the highly dynamic nature of the environment, they are constantly changing. This presents an issue when implementing GIS in coastal geomorphology [12; 9]. Raper [12] recognized that the specification of a formal or informal ontology could help to overcome this problem.

Ontology depends on the model-maker and the context and may vary for different users [9]. Thus, it is important to define the purpose and intended users of an ontology prior to creation. An informal ontology for reef islands, which would facilitate mutual understanding between researchers, managers and local communities, is desirable. Once formalized, such an ontology could also aid the application of information technologies to the forecasting and monitoring of climate-change-related impacts in reef island environments [13].

This paper represents the first step towards achieving this goal. It develops a small, prototypical reef island domain ontology, based on informal, natural language relations, for 20 fundamental terms within the domain. A subset of these terms and their relations is discussed in detail. Alignment with the upper-level DOLCE ontology are also discussed. While this work does not achieve a complete or workable ontology for the domain yet, it reveals and discusses challenges which must be overcome in the creation of a reef island domain ontology and which could be relevant to other ontologies in the geospatial domain.

1.2   Previous Studies

Despite the obvious benefits, a domain specific reef island ontology does not yet exist and, to my knowledge, had not been attempted prior to this study. Though a number of studies have applied ontologies in coastal environments.

For example, Moore *et al.* [14] demonstrated how ontologies could be employed to facilitate Integrated Coastal Zone Management. Van de Vlag *et al* [9] employed ontologies for the identification of beaches requiring nourishment. Their work focused on the development of a "problem" and a "product" ontology which were based on data available for beach objects. The authors found that physical processes can provide a framework for an ontology in natural systems.

Myers *et al* [13] were the first to apply semantics and ontology to the study of coral reefs. They formulated an ecosystem ontology to make unconnected sensor data sources computer-understandable to enable an early warning system for coral bleaching.

2   Methodology

No single, widely accepted, method for the creation of a domain ontology exists [15]. It is widely accepted however, that in order to create a coherent, systematic and complete ontology it should be aligned to a higher level, foundational (upper-level) ontology which defines important overarching components of the earth and the relationships between them [16]. Frank [17] states that, within the context of Geographic Information Science (GIS), ontologies should include space, time, objects and processes as defined in an upper-level ontology.

The ontology presented here will be aligned to DOLCE (Descriptive Ontology for Linguistic and Cognitive Engineering) as the top level ontology [18]. DOLCE aims to "negotiate meaning" of terms/entities at a foundational level which will enable co-operation and consensus between humans and "artificial agents" [19]. The basic categories of DOLCE, relevant to this application, are presented in Figure 2.

A 'top down' / combination approach will be employed to formulate the ontology. The most important terms and concepts in the reef island domain will be determined and listed. These terms will be defined and classified. The relationships between the terms will be defined in natural language (Table 2).

It was decided that, for the purposes of this paper, a preliminary ontology incorporating just 20 key terms from the reef island domain would be developed in natural language. As well as providing experience in the creation of ontologies, and demonstrating the complexity of the reef island domain, this will unveil the challenges associated with the creation of an ontology for the domain and provide insight into the usefulness of ontologies within the domain.

Table 1. Fundamental particulars of the Reef Island domain to be included in the preliminary domain ontology. DOLCE Basic Categories are also specified.

Physical Endurants	Perdurants
Reef *(POB)*	Coral Growth *(PRO)*
Reef Island *(POB)*	Sediment Production *(PRO)*
Wave *(POB)*	Sediment Transport *(PRO)*
Current *(POB)*	Erosion *(EV, PRO)*
Wind *(POB)*	Accretion *(EV, PRO)*
Sediment *(M)*	Inundation *(EV, PRO)*
Coral *(POB, M)*	Sea Level Rise *(PRO)*
Beach *(F/POB)*	
Harbor *(POB)*	
Sand bar *(POB)*	
Spit *(POB)*	
Dune *(POB)*	
Vegetation *(POB)*	

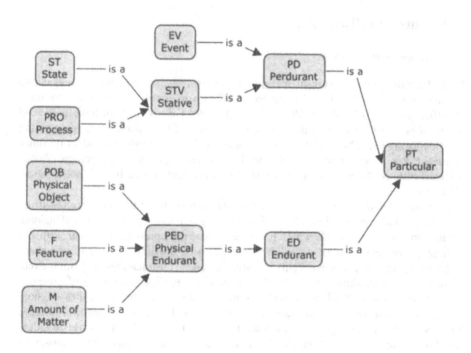

Fig. 2. Relevant ontological categories of the DOLCE foundational ontology. The reef island domain ontology will be aligned to these classes (source: 18).

Twenty particulars deemed to be of the greatest importance to the definition of reef islands and their sustainable management were chosen. The relationships of importance to management issues such as erosion, accretion and inundation were included with the thought to, in future, create an automated early warning/prediction system.

These twenty terms are listed in Table 1 and have been broadly divided into *physical endurants* and *perdurants*. In accordance with the DOLCE upper level ontology *endurants* are entities which are present in full at any time that they are present, while *perdurants* are processes which extend through time by accumulating different temporal parts [19]. Thus, *perdurants* are only partially present at any time as their past and future "parts" are not present at all times [19].

Physical endurants have a clear spatial location and are divided within the DOLCE Ontology into 3 basic categories: *Physical Objects (POB), Amounts of Matter (M)* and *Features (F)* [18]. Two basic categories of *perdurants*, *Process (PRO)* and *Event (EV)*, are also distinguished here. Refer to sections 3.1 and 3.4 for discussion of the meaning of these categories and their application to the reef island domain particulars mentioned here.

A subset of the most important of these terms and their relations will be discussed in more detail (Fig. 4). The terms chosen for detailed discussion are reef flat, reef island, sediment, sediment production, coral and beach.

3   Results and Discussion

3.1   Physical Endurants

Table 1 outlined the division of the chosen *physical endurants* from the reef island domain into three basic DOLCE categories shown in Figure 2. The majority were classified as *physical objects (POB)* which, by DOLCE's definition, have unity and *temporal parts* (meaning that they can change some of their parts while maintaining their identity) [18]. Their existence is not specifically, constantly dependent on other objects. For instance, a reef island may have a sand spit as it's proper part but if it looses this spit to erosion the reef island is still a reef island. It does not loose its identity.

Amounts of matter (M) are those *endurants* with no unity that are referred to by mass nouns like "gold" [18]. DOLCE recognizes a*mounts of matter* as *extensional entities*, meaning, all entities which have the same proper parts are identical [20]. For example, every entity called "sand" has sand grains as its only proper part and every entity composed only of sand grains is called "sand". Thus, mass nouns from the reef island domain, including coral and sediment, can be classified as *amounts of matter*.

Features (F) in DOLCE are essentially whole entities with unity but they are constantly dependent on *physical objects* as their hosts. Examples of *features* include holes, surfaces or stains [18]. If a feature is removed from its host it looses its identity. For example, if a "red wine stain" is no longer on a shirt (its host) it looses its identity as a "stain" and is simply some red wine. Transversely, a body part, like a "hand", is NOT considered a *feature* as, even when it is not attached to a body, it is still recognizable as a hand [18].

In the reef island domain it is interesting to apply this example to a "beach". A "beach" is defined to a large extent by its location on the land at the boundary with the sea. If a beach is taken away from the coast would it maintain its identity or would it simply be a "body of sand"? If so, it could be classed as a *feature* in DOLCE.

3.2 Relations

Figure 3 presents the natural language relations between the 20 reef island particulars listed in Table 1. These relations are briefly and informally defined in Table 2. The main purpose of Figure is to demonstrate the complexity of the domain even when dealing with just a few entities. Thus, it will not be discussed in detail. This paper will discuss in detail the relations between the subset of these terms presented in Figure 4.

Table 2. Natural language descriptions of the relationships used in the extended prototype reef island ontology (Fig. 3)

Relationship	Definition
Is a	Every instance of A is instance of B, e.g. "dog" IS_A "mammal".
Part of	Refer to discussion of parthood below.
Quality of	Attribute or characteristic of an entity which can be measured or described. e.g. the weight of a pen [19] or the height of a wave.
Participates in	Is involved in an occurence/process (*pedurant*) [18].
Instance of	An example of. e.g. Warraber Island is an instance of Reef Island.
Occurs on	Used for processes and entities, implying where, or on what, a process or entity is found or works. e.g. the process of erosion occurs on the beach.
Intersects (cross)	Meet at a point.
Leads to	Refers to a process that either by itself or in combination with something else causes a phenomena. e.g. lack of food leads to hunger.

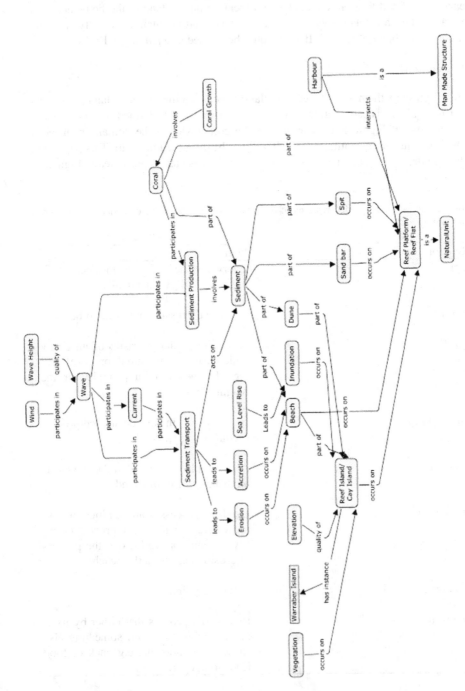

Fig. 3. Prototype, natural language ontology for 20 key particulars in the reef island domain

3.3 Parthood

The parthood relation is one of the primitive relations of Mereotopology for Individuals (MI) presented by Donnelly and Bittner, [21] and is fundamental to the subset ontology presented here (Figure 4). Two variations, or levels, of the *part_of* relation exist in this preliminary reef island ontology. They are described, in accordance with Donnelly and Bittner [21], as *Part* $_{all-1}$ and *Part* $_{all-2}$.

Part $_{all-1}$ is the relation between two classes if and only if every instance of class A is a part of some instance of class B. For instance, every reef island is part of a reef flat but not every reef flat has a reef island as its part (Figure 4). Thus, the relationship *Part* $_{all-1}$ (Reef Island, Reef Flat) holds in the reef island domain. A reef island and reef flat overlap such that the region occupied by the reef island is completely contained within the region occupied by the reef flat (refer back to Figure 1). Thus, location relations such as *overlap* and *coincide* (see 21) also reaffirm the *part_of* relation.

Part $_{all-2}$ is the relation between class A and B if and only if every instance of B has some instance of A as its part. For example, every reef flat has coral as its part but not every coral is part of a reef flat. Similarly, every reef island has a beach as a part but not every beach is part of a reef island; and every beach has sediment as a part but not all sediment is part of a beach. Thus, the relations

Part $_{all-2}$ (Coral, Reef Flat); *Part* $_{all-2}$ (Beach, Reef Island) and *Part* $_{all-2}$ (Sediment, Beach) all hold (Figure 4).

The ontology presented here deals with synonyms by including the synonymous terms in the same class. *Instances* are deemed to be one specific or particular member of a class. For example, Warraber Reef Island in the Torres Strait, Australia is an *instance* of the class reef island.

3.4 Perdurants

Perdurants (occurrences) are divided in DOLCE into, *eventives (EV)* or *statives (STV)* based on whether they are cumulative (Fig. 2). *Statives* are further divided into *states (ST)* and *processes (PRO)* (Refer to 20 for further explanation). Table 1 shows the division of some *perdurants* from the reef island domain into *processes* and *events*. Erosion, accretion and inundation can be classified within both these categories. For example a beach can be *eroding (PRO)* during a storm and that storm causes an *erosion event (EV)*. Coral growth, sea level rise, sediment production and transport were all classed as *processes* as they refer to continuous processes with parts which are not all present simultaneously. For example, "sediment production at low tide is more efficient than sediment production at high tide". At least two parts of the sediment production process exist that are not present at the same time.

The relations between *endurants* and *perdurants* (processes) in the reef island domain are difficult to define. The most common relation between *endurants* and *perdurants* defined in the DOLCE is that of *participation* [19; 18]. *Participation* is a time regular relation between *endurants* and *perdurants* [22]. A *perdurant* could not occur if it was not for the involvement of some *endurant(s)* and the *endurant* "lives" in time by participating in some *perdurant(s)*. For example, a runner *(endurant) participates* in a race (which is a *perdurant* as it has a start middle and end that are not all present at one time) [18].

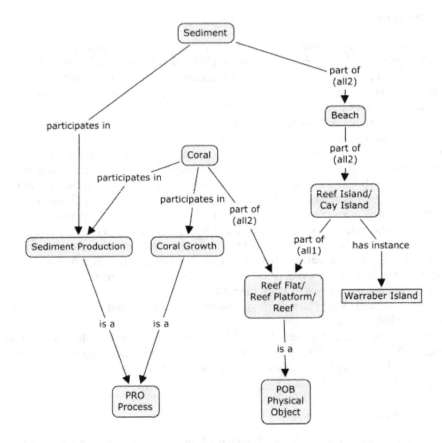

Fig. 4. Ontological relations between a selected subset of particulars from the reef island domain. These classes align with the classes of the upper-level DOLCE ontology shown in Fig. 2.

In a reef island context the *participation* relation could be used to describe the link between corals and the process of sediment production. Sediment is produced by waves breaking up corals. Thus, waves, corals and sediment *participate* in sediment production. However, these participants obviously play very different roles in the process. The waves are the agents, the corals are the source and the sediment is the result of the process. However, it is not obvious within DOLCE how these subtleties can be defined.

3.5 Challenges and Future Work

The complex, dynamic and inter-linked processes which occur on, and form, reef islands make the definition of formal and explicit boundaries and relationships challenging [23, 24]. It is particularly difficult to describe the relationships between *perdurants* (such as sediment production, sediment transport and erosion) and between *endurants* and *perdurants* (such as sediment and erosion).

Feedback loops, exist between reef flat geomorphology and waves and currents whereby the reef flat morphology alters the wave patterns occurring on it but its own morphology is also controlled in a large part by these processes. Reef Islands are entirely dependent on the surrounding reef and thus the coral and other organisms that are part of the reef. As coral is dependent on living zooxanthellae - who's survival is governed by numerous factors including water quality, availability of light and nutrients etc. - they are highly vulnerable to environmental conditions. The nominal ontology presented here falls short of incorporating these dependencies and vulnerabilities. Such intricacies are fundamental to creating a useful domain ontology and defining processes of erosion, inundation and sea level rise which are perhaps of the greatest importance from a management perspective. Future work in this direction is required.

As Reef Island processes are exceedingly complex and difficult to describe using highly expressive natural language their description in less expressive formal languages is even more challenging. It is further complicated for some entities by the use of the same term as a noun and a mass noun. For instance, "coral" is used in natural language in a number of different ways - "the coral that makes up the reef" (referring to the collective mass noun, numerous corals of different species composing a single reef flat) "this reef is composed mainly of *Porites annae* coral" (referring to a species of coral) and "there is a clam on this coral" (referring to a particular organism). These subtleties are difficult to make explicit in a formal ontology.

Further exploration of the spatiotemporal co-location properties of DOLCE [18] is needed to explicate relations like "during its life, the beach *(POB)* is composed of sediment *(M)* so these are spatiotemporally co-localised". The addition of *qualities* to the nominal ontology presented here is also necessary in the future particularly if the ontology is to be employed for prediction purposes as in Myers *et al.* [13].

4 Conclusions

The creation of a reef island domain ontology is highly desirable to improve the understanding of reef island processes, allow better communication between different stake holders and ensure interoperability between data sources. This paper presents an initial step towards the development of an ontology for the reef island domain. It discusses challenges presented by the complexity, dynamism and interlinkedness of processes and entities in the domain. The findings and relations discussed in this paper provide insight to the further development of the reef island domain ontology and could also be useful in the creation of ontologies for other dynamic geographic domains.

Acknowledgments. The author would like to acknowledge and thank the Institute for Geoinformatics (IFGI), University Muenster, Germany for their generous financial and academic support.

References

1. Gourlay, M.R.: Coral Cays: Products of Wave Action and Geological Processes in a Biogenic Environment. In: Proceedings Sixth International Coral Reef Symposium, Townsville, August 8 -12 (1988)
2. Yamano, H., Miyajima, T., Koike, I.: Importance of Foraminifera for the Formation and Maintenance of a Coral Sand Cay: Green Island, the Great Barrier Reef, Coral Reefs, Australia, vol. 19, pp. 51–58 (2000)
3. Hopley, D.: Sediment Movement Around a Coral Cay, Great Barrier Reef, Australia. Pacific Geology 15, 17–36 (1981)
4. Nicholls, R.J., Wong, P.P., Burkett, V.R., Codignotto, J.O., Hay, J.E., McLean, R.F., Ragoonaden, S., Woodroffe, C.D.: Coastal Systems and Low-Lying Areas. In: Parry, M.L., Canziani, O.F., Palutikof, J.P., van der Linden, P.J., Hanson, C.E. (eds.) Climate Change 2007: Impacts, Adaptation and Vulnerability. Contribution of Working Group II to the Fourth Assessment Report of the Intergovernmental Panel on Climate Change, pp. 315–356. Cambridge University Press, Cambridge (2007)
5. Guarino, N.: Formal Ontology in Information Systems. In: Guarino, N. (ed.) Formal ontology in Information Systems. Proceedings of FOIS 1998, Trento, Italy, June 6-8 (1998)
6. Gruber, T.R.: A Translation Approach to Portable Ontology Specifications. Knowledge Systems Laboratory. Technical Report KSL 92.71, Stanford University (1993)
7. Bishr, Y.: Overcoming the semantic and other barriers to gis interoperability. International Journal of Geographical Information Science 12(4), 299–314 (1998)
8. Noy, N.F., McGuinness, D.L.: Ontology Development 101: A Guide to Creating Your First Ontology. Stanford Knowledge Systems Laboratory Technical Report KSL-01-05 (2001), http://www-ksl.stanford.edu/people/dlm/papers/ontology-tutorial-noy-mcguinness-abstract.html (accessed: 28-06-09)
9. Van de Vlag, D., Vasseur, B., Stein, A., Jeansoulin, R.: An Application of Problem and Product Ontologies for the Revision of Beach Nourishments. International Journal of Geographical Information Science 19(10), 1057–1072 (2005)
10. Chandrasekaran, B., Josephson, J.R., Benjamins, V.R.: What Are Ontologies, and Why Do We Need Them? IEEE Intelligent Systems 14(1), 20–26 (1999)
11. Hopley, D., Smithers, S.G., Parnell, K.E.: The Geomorphology of the Great Barrier Reef: Development, Diversity and Change. Cambridge University Press, Cambridge (2007)
12. Raper, J.: Multi Dimensional GIS: Extending GIS in Space and Time, p. 320. Taylor and Francis, London (1999)
13. Myers, T.S., Atkinson, I., Johnstone, R.: Supporting Coral Reef Ecosystems Research through Modeling Re-usable Ontologies. In: Meyer, T., Orgun, M.A. (eds.) Proc. Knowledge Representation Ontology Workshop (KROW 2008), Sydney, Australia. CRPIT. ACS, vol. 90, pp. 51–59 (2008)
14. Moore, A., Jones, A., Sims, P., Blackwell, G.: Integrated Coastal Zone Management's Holistic Agency: An Ontology of Geography and GeoComputation. Presented at the 13th Annual Colloquium of the Spatial Information Research Centre, University of Otago, Dunedin, New Zealand (2001)
15. Gomez-Perez, A., Fernandez, M., De Vicente, A.: Towards a Method to Conceptualize Domain Ontologies. In: Workshop on Ontological Engineering, ECAI 1996, Budapest, Hungary (1996)
16. Bittner, T.: Ontology-Based Framework for the Analysis and Integration of Ecoregion Classification Systems. Journal of Geographic Information Science, 1–19 (in Press)

17. Frank, A.: A linguistically justifiEd proposal for a spatiotemporal ontology. In: Proceedings of the COSIT 2003 international conference, Position paper in COSIT 2003 workshop (2003)
18. Masolo, C., Borgo, S., Gangemi, A., Guarino, N., Oltramari, A.: Ontology Library. Deliverable D18, ISTC-CNR (2003)
19. Gangemi, A., Guarino, N., Masolo, C., Oltramari, A., Schneider, L.: Sweetening Ontologies with DOLCE. In: Gómez-Pérez, A., Benjamins, V.R. (eds.) EKAW 2002. LNCS (LNAI), vol. 2473, p. 166. Springer, Heidelberg (2002)
20. Gangemi, A., Guarino, N., Masolo, C., Oltramari, A.: Understanding Top-Level Ontological Distinctions. In: Gomez Perez, A., Gruninger, M., Stuckenschmidt, H., Uschold, M. (eds.) Proceedings of the IJCAI 2001 Workshop on Ontologies and Information Sharing, Seattle USA, pp. 28–34 (2001)
21. Donnelly, M., Bittner, T.: Spatial relations between classes of individuals. In: Cohn, A.G., Mark, D.M. (eds.) COSIT 2005. LNCS, vol. 3693, pp. 182–199. Springer, Heidelberg (2005)
22. Borgo, S., Masolo, C.: Foundational Choices in DOLCE. In: Staab, S., Studer, R. (eds.) Handbook on Ontologies, 2nd edn., pp. 361–382. Springer, Heidelberg (2009)
23. Preobrazhensky, B.V.: Problems of Studying Coral-Reef Ecosystems. Heloglaender wiss. Meeresunters 30, 357–361 (1977)
24. Insalaco, E., Skelton, P., Palmer, T.J.: Carbonate Platform Systems: Components and Interactions – An Introduction. Geological Society 178, 1–8 (2000)

Narrative Geospatial Knowledge in Ethnographies: Representation and Reasoning*

Chin–Lung Chang[1,3], Yi–Hong Chang[2], Tyng–Ruey Chuang[1],
Dong–Po Deng[1,4], and Andrea Wei–Ching Huang[1]

[1] Institute of Information Science
[2] Institute of Ethnology
Academia Sinica
Taipei, Taiwan
[3] Department of Computer Science and Information Engineering
National Taiwan University of Science and Technology
Taipei, Taiwan
[4] ITC — International Institute for Geo–Information Science and Earth Observation
Enschede, The Netherlands

Abstract. Narrative descriptions about populated places are very common in ethnographies. In old articles and books on the migration history of Taiwan aborigines, for example, narrative sentences are the norms for describing the locations of aboriginal settlements. These narratives constitute a form of geospatial knowledge, and there is a need to develop knowledge representation and reasoning techniques to help analyze literatures, and to aid field works. In this paper, we outline the design of a formal vocabulary to represent and reason about geospatial narratives about populated places, keeping as close as possible to the phrases used in ethnographies. The vocabulary is implemented as OWL concepts and properties, and the rules for geospatial reasoning are expressed in SWRL.

1 Narrative Geospatial Knowledge

In research and study about People and Place, it is necessary to acquire and analyze geospatial information about populated places. Such geospatial information — the location of a settlement relative to certain geographic features and other settlements, for example — often is described in a natural language, and the descriptions themselves are expressed in an everyday vocabulary that is intuitive to human but difficult to process automatically. As an example, let us look at the following sentence, which is taken and translated from an ethnohistorical article about the Atayal aborigine in Taiwan:

> Kanaongan Settlement ... is located at the right bank of Dacingshuei River, where the river meets the ocean. [17, p. 137]

---

* This research is supported in part by National Science Council of Taiwan under grant NSC 98-2410-H-001-075-MY2 ("Ontology–based Collaborative Production of Geospatial Information").

K. Janowicz, M. Raubal, and S. Levashkin (Eds.): GeoS 2009, LNCS 5892, pp. 188–203, 2009.

The location of Kanaongan Settlement is described by two statements about the place's geospatial relations to others: 1) it is at the right bank of the Dacingshuei river, and 2) it is near the Dacingshuei estuary. Although we are not given the coordinates of the Kanaongan settlement, we have a good idea of the settlement's position, even more so if we consider the facts that the Dacingshuei river is eastbound, and it flows into the Pacific Ocean. That is, we can conclude that the settlement is near by the southern river bank of the Dacingshuei estuary (as the right bank of an eastbound river is the southern bank of the river).[1] Suppose that we learn from the same article or other sources that settlement X and Kanaongan Settlement are on the opposite sides of the same river, and they face each other. We will be able to conclude that settlement X is near by the northern river bank of the Dacingshuei estuary. Note that the conclusion is drawn without the need to know the location of the Dacingshuei estuary.

Narrative descriptions about geospatial relations among Places — collected from ethnohistorical records or field interviews — constitute a kind of geospatial knowledge that is rich in domain semantics, difficult to acquire by other means, and defying easy assimilation and analysis in and by conventional geographic information systems (GIS). To systematically process and analyze large collections of these geospatial narratives, therefore, it calls for new methods and new techniques. If successful such developments will help shed new lights on several areas such as knowledge representation and reasoning, ontology and semantic web, humanistic GIS, etc.

We list below several characteristics of narrative geospatial descriptions. These characteristics also magnify themselves as the main technical issues in processing narrative geospatial knowledge.

- The narratives are expressed in everyday vocabularies that are rich (even diverse) in their linguistic and cultural interpretations (the "left bank" of a river, the "foot" of a mountain, etc.).
- Directional and relative terms are used to describe the location of a place, as well as its spatial relations with others. The terms are approximative ("opposite to", "about 100 kilometers away", "near by", etc.).
- Places are identified by (common) names, not by coordinates. Their positions and footprints are left unspecified. If specified, they are imprecise or vague by definition. (For example, what is the definite spatial extent of a mountain?)
- The descriptions about a place can be conflicting, incomplete, or missing. Often such places are of an ethnohistorical nature and cannot be identified by nowadays technologies. Nevertheless they can be used as the primary references for other places. (Where is Shangri–La?).

---

[1] However, the map in the same article puts Kanaongan Settlement at the northern bank of the Dacingshuei estuary [17, p. 135]. (Also see Figure 6 in Appendix A.) This case illustrates that geospatial reasoning can be used to detect inconsistency in geospatial statements — in the article either the sentence or the map is wrong. Or it could be that our understanding of the term "right bank" is different from that of the author.

2   A Vocabulary for Qualitative Geospatial Expressions

The main purpose of qualitative spatial representation and reasoning is to make explicit common–sense knowledge, so that given appropriate reasoning techniques, a computer could make prediction, diagnose and explain the behavior of physical system in qualitative manner without resolving to an often intractable or perhaps unavailable quantitative model [6]. Therefore, qualitative spatial representation and reasoning not only acts as a model to clarify formal semantics of qualitative spatial objects and relations from narrative descriptions, but is also used to find new information from what is already known. There are numerous studies on this subject, on various aspects of spatial relationships such as topology, orientation, distance, size, and shape [4,5,9]. However, given the nature of geospatial information present in narrative descriptions, currently we only concern about the representation of certain topological, directional, and orientational information when specifying qualitative relations among geospatial entities.

As a study on narrative geospatial knowledge, we have looked into ethnographies for descriptions about the locations of Taiwan aboriginal settlements, so as to use them as the sources of actual vocabularies for geospatial expressions. Take the following sentences about Sikilian Settlement as examples.[2]

> Sikilian Settlement is located at the left bank of midstream Liwu River, and is about 1.6 kilometers to the northwest of the junction of its branch Wahei-er River. It is opposite to the small terrace slightly east of and below the Syuejiachang Station on Central Cross–Island Road. That is, it is at the mountain belly north of Mantou Mountain. [16, p. 178]

One immediately notices geospatial phrases that may subject to different interpretations ("left bank" and "mountain belly"). The use of size and direction modifiers are also problematic ("the *small* terrace *slightly* east of"). There are pronouns and missing nouns to resolve too ("the junction of *its* branch Wahei-er River"). Nevertheless, from these sources we have identified a small set of phrases frequently used for geospatial references:

- phrases for directional references such "is *x* kilometers to the northwest of" and "is north of";
- phrases for orientational references such as "left bank" and "opposite to";
- phrases for references to a part of an geospatial object, such as the "midstream" of a river, and the "belly" of a mountain;
- phrases for different types of natural features ("river" and "junction") and artificial landmarks ("road" and "station").

We list in Appendix A a set of nine sample paragraphs drawn from the same ethnography [16]. They are all about the locations of (historical) aboriginal settlements. Based on the geospatial narratives used in these paragraphs, we develop

---

[2] These are direct quotations, but of our translations. The original texts are written in traditional Chinese. We take care in making accurate translations of the narratives, in particular about how geospatial references are used in the original texts.

Author Index

a vocabulary for geospatial references, keeping as close to the phrases that are actually used in these paragraphs. Our goal is that, by using this vocabulary, we can represent and reason in a formal way the geospatial narratives in these ethnographies.

Note that we are not aiming at new techniques which will automatically process natural language texts for the extraction of geospatial knowledge. Rather, we seek systematic methods to express geospatial knowledge in ethnographic narratives so that such knowledge can be aggregated and analyzed, and becomes more useful and reusable to human and machines. In this paper, we only deal with qualitative geospatial knowledge; we consider direction but not distance, for example. We have also avoided relying on numerical calculation in the representation and reasoning of geospatial narratives.

3   Framing a Geospatial Knowledge Domain

Before setting out to acquire and process narrative geospatial knowledge, we need to decide on the scope and the level of details of narrative geospatial knowledge when it is represented in a system. The decision inevitably depends on the formalism we are adapting, and is constrained by the restrictions of the formalism. For this case study, we have settled on OWL DL (the Description Logic subset of the OWL Web Ontology Language) [1] as the representation formalism for the narrative knowledge, and use SWRL (a Semantic Web rule language combining OWL and RuleML) [2] for rule representation and inference.

Compared to the decision on a representation mechanism (OWL DL in our case), the process of adapting and/or devising an appropriate ontology, and that of mapping the assumptions and vocabularies in a narrative knowledge domain to those in the formal ontology, is much harder. These decisions frame the domain of the narrative knowledge, hence, effect the kinds of facts to be admitted and to be reasoned with in the system. Let us take the term "mountain" as an example. In our study of geospatial narratives, "mountain" almost always refers to the area of a mountain. The terms "top", "side" and "foot" are often used in combination with term "mountain" to refer to the top, side, and foot sub–areas of a mountain area, respectively. These areas and sub–areas may be inhabited by people. In this section, we will use this example to elaborate some of the considerations when framing the domain of geospatial areas, and use these areas as references to the places of settlements.

One can choose to use Region Connection Calculus (RCC) [5] to model the various mountain sub–areas and their relations. An area is mapped to a region in the sense of RCC, and the relation between two areas is but one of the RCC–8 (or rather, RCC–5) relations. That is, each of the mountain top, mountain side, and mountain foot regions *is a proper part of* (PP) the mountain region, see Figure 1. Figure 1, however, expresses more. It illustrates that the mountain region is partitioned into three sub–regions of which the mountain side region surrounds the mountain top region, and the mountain foot region surrounds the mountain side region. Observing this, it could be useful to first define three regions m, s, t of which

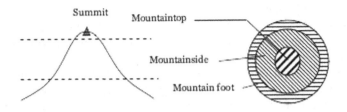

Fig. 1. Topological relations among the top, side, and foot areas of a mountain

m is the region for the entire mountain area, and PP(s, m) and PP(t, s). One can then describe the mountain top area as t, the mountain side area as the area of $s - t$, and the mountain foot area as $m - s$. However, in our case study geospatial narratives do not impose or require topological precision of this kind. The use of a region difference operator on top of RCC–5, we think, will also impose further demands to a system based on OWL DL and SWRL.

One can choose instead to model the top, side, and foot sub–areas of a mountain as three distinct parts of a mountain area, but without explicit mentioning their topological relations. That is, a mountain area will have three separate object properties, each keeping the top, the side, and the foot sub–area, respectively. The classes for the three kinds of sub–areas, as well as that of the entire mountain areas, moreover, are kept distinct from one other, so there will be no confusion of mistaking one kind for another kind. In description logic, they are disjoint concepts. We have adapted this approach to modeling inhabited areas that are identified by common geographic names. That is, in our narrative geospatial domain, the part–whole relationship of a place is elaborated more than its topological relationship.

For the area (and each of their sub–areas) identified by a common geographic name (e.g., "the foot of mountain X"), we require that it also possesses a "position" property indicating its geophysical location. For example, for the top, the side, and the foot areas of a mountain (and the mountain itself), they all have the summit as their position. By their unique position, the sub–areas can be related together, and associated to the mountain to which they belong. Here, a position serves two purposes — as the coordinates to be used when the geographic feature is visualized on a map, and as an (imprecise) identifier from which parts of the feature can be related together and associated with the whole. Currently we do not require an area to have a "footprint" (or, boundary) property, as such property is not easily quantifiable, even in non–narrative geospatial domains.

By modeling areas referred to by common geographic names as above, we establish a framework of geospatial references for settlements, whose positions can be vague or lost. By translating narrative descriptions of settlement locations to directional, orientational, and topographical relations to existing (and persistent) geographic features and their positions, in our model, we therefore establish the relative positions of the settlements.

4   Representation and Reasoning

We first give a short introduction to an alternative syntax we devise for OWL DL, then proceed to describe our current implementation for the representation and reasoning of geospatial narratives in ethnographies. Figures 2 and 3 list some of the basic concepts and common properties we use in the implementation. The syntax we use is very close to those used in Description Logic (or OWL DL) but is more concise.

Named concepts are introduced by the keyword Class. An anonymous concept can be constructed from a property with certain restrictions on its range, *e.g.*, "originateFrom min 1" and "meet some Upstream". In Figure 2, for example, a River is defined as a Place that originates from at least one Place, ends at one Place, has one Upstream, has one Midstream, and has one Downstream. (From now on, we will simply use the term 'a river' to refer to an individual/object in the concept/class River.) Upstream, Midstream, and Downstream, furthermore, are all subclasses of Place. An Upstream has at least one position (of class Point), a Downstream has one position (of class Point), and a Midstream must meet some Upstream and some Downstream. The intuition behind such definitions is that the place from which a river originates are associated to the sources of its upstream, and the place from which a river ends is associated to the sink of its downstream. Moreover, the up/middle/down streams of a river are pair–wise connected by the 'meet' relations as required in the definition of Midstream.

Figure 3 shows common properties about places; these are relations among the classes we just define. The keyword Property declares a relation between two classes; it specifies the name of the relation, and the class names of its domain and its range. The square bracket immediately follows the declaration can specify the property to be "functional", "inverseFunctional", "transitive", and/or "symmetric" if that is the case for the property. An inverse property f, if exists, is indicated at the end with the notation f^{-1}. In Figure 3, for example, it declares that "southOf" is a transitive property of places with "northOf" as its inverse property.

By using SWRL, we can define additional properties by composing them from existing properties. See Figure 4 for some basic rules for geospatial reasoning. This way, we are free from asserting in OWL DL those facts that can be deduced from SWRL rules, *e.g.*, the fact that a is northeast of b given the facts that a is east of some individual x, and x is north of b. The rules in Figure 4 basically fall into two categories. The first category is about the transfer of property from the part to the whole. For example, if x is located at a mountaintop t which is belonging to a mountain m, then x is located at mountain m as well. The second category is abut common–sense geospatial reasoning such as the northeast rule we just mentioned.

In our case, the geospatial narratives used in ethnographies are very concise, but to define the corresponding common–sense geospatial reasoning rules can be quite involving. For example, given the following geospatial narratives:

Class	People.	
Class	Place.	
Class	Settlement	$\sqsubseteq$ Place $\sqcap$ (populatedBy **min** 1).

Class	Geo.	
Class	Point	$\equiv$ Geo $\sqcap$ (hasHeight **exactly** 1) $\sqcap$
		(hasLatitude **exactly** 1) $\sqcap$ (hasLongitude **exactly** 1).
Class	Landmark	$\equiv$ Place $\sqcap$ (hasPosition **exactly** 1).
Class	Bridge	$\sqsubseteq$ Landmark.
Class	Station	$\sqsubseteq$ Landmark.

Class	Mountain	$\equiv$ Place $\sqcap$ (hasPosition **exactly** 1) $\sqcap$ (hasTop **exactly** 1) $\sqcap$
		(hasSide **exactly** 1) $\sqcap$ (hasFoot **exactly** 1).
Class	Mountaintop	$\sqsubseteq$ Place $\sqcap$ (hasPosition **exactly** 1).
Class	Mountainside	$\sqsubseteq$ Place $\sqcap$ (hasPosition **exactly** 1).
Class	Mountainfoot	$\sqsubseteq$ Place $\sqcap$ (hasPosition **exactly** 1).
Disjoint		Mountaintop, Mountainside, Mountainfoot.

Class	River	$\equiv$ Place $\sqcap$ (originateFrom **min** 1) $\sqcap$ (endAt **exactly** 1) $\sqcap$
		(hasUpstream **exactly** 1) $\sqcap$ (hasMidstream **exactly** 1) $\sqcap$
		(hasDownstream **exactly** 1).
Class	Upstream	$\sqsubseteq$ Place $\sqcap$ (hasPosition **min** 1).
Class	Downstream	$\sqsubseteq$ Place $\sqcap$ (hasPosition **exactly** 1).
Class	Midstream	$\sqsubseteq$ Place $\sqcap$ (meet **some** Upstream) $\sqcap$ (meet **some** Downstream).
Disjoint		Upstream, Midstream, Downstream.

Class	RiverCross	$\sqsubseteq$ Place $\sqcap$ (hasPosition **exactly** 1) $\sqcap$ (hasInflow **min** 2)
Class	PlaceGroup	$\sqsubseteq$ Place.
Class	Pair	$\sqsubseteq$ PlaceGroup $\sqcap$ consistOf **exactly** 2.

Fig. 2. Concepts for places and settlement in ethnographies

Sakahen Settlement is opposite to Basawan Settlement; they are separated by Mugua River. Basawan is located at the right bank of upstream Mugua River.

we can infer that Sakahen is located at the left bank of the Mugua river. For this, we need to formulate and use the following rule:

$$\text{Pair}(p) \wedge \text{consistOf}(p, x) \wedge \text{consistOf}(p, y) \wedge \text{differentFrom}(x, y) \wedge$$
$$\text{River}(r) \wedge \text{inBetween}(r, p) \wedge \text{rightBankOf}(x, r) \rightarrow \text{leftBankOf}(y, r).$$

Before the above rule can be applied, we will first need to pack the two settlements into a pair, as well as to assert that the Mugua river is in between the pair, in addition to the usual assertions that Basawan is a settlement, Mugua is a river, *etc.*

We also need rules to enforce integrity constraints governing the facts admitted to a geospatial knowledge base. Earlier we mentioned the intuition behind

Property populatedBy(Settlement, People).
Property hasPosition(Place, Point).
Property locatedAt(Place, Place).

Property hasTop(Place, Mountaintop).
Property hasSide(Place, Mountainside).
Property hasFoot(Place, Mountainfoot).

Property originateFrom(River, Place).
Property endAt(River, Place).
Property hasUpstream(Place, Upstream).
Property hasMidstream(Place, Midstream).
Property hasDownstream(Place, Downstream).
Property meet(Place, Place)   [symmetric].

Property bankOf(Place, River).
Property leftBankOf(Place, River)   ⊑   bankOf.
Property rightBankOf(Place, River)   ⊑   bankOf.

Property hasInflow(RiverCross, River).
Property consistOf(PlaceGroup, Place).
Property inBetween(Place, Pair)   separatedBy^{-1}.
Property separatedBy(Pair, Place)   inBetween^{-1}.

Property southOf(Place, Place)   [transitive]   northOf^{-1}.
Property northOf(Place, Place)   [transitive]   southOf^{-1}.
Property westOf(Place, Place)   [transitive]   eastOf^{-1}.
Property eastOf(Place, Place)   [transitive]   westOf^{-1}.
Property southWestOf(Place, Place)   [transitive]   northEastOf^{-1}.
Property northEastOf(Place, Place)   [transitive]   southWestOf^{-1}.
Property southEastOf(Place, Place)   [transitive]   northWestOf^{-1}.
Property northWestOf(Place, Place)   [transitive]   southEastOf^{-1}.

Fig. 3. Common properties about places

$$\texttt{locatedAt}(s, top) \land \texttt{hasTop}(m, top) \rightarrow \texttt{locatedAt}(s, m).$$
$$\texttt{locatedAt}(s, side) \land \texttt{hasSide}(m, side) \rightarrow \texttt{locatedAt}(s, m).$$
$$\texttt{locatedAt}(s, foot) \land \texttt{hasFoot}(m, foot) \rightarrow \texttt{locatedAt}(s, m).$$

$$\texttt{eastOf}(s, top) \land \texttt{hasTop}(m, top) \rightarrow \texttt{eastOf}(s, m).$$
$$\texttt{eastOf}(s, side) \land \texttt{hasSide}(m, side) \rightarrow \texttt{eastOf}(s, m).$$
$$\texttt{eastOf}(s, foot) \land \texttt{hasFoot}(m, foot) \rightarrow \texttt{eastOf}(s, m).$$
$$\texttt{westOf}(s, top) \land \texttt{hasTop}(m, top) \rightarrow \texttt{westOf}(s, m).$$
$$\texttt{westOf}(s, side) \land \texttt{hasSide}(m, side) \rightarrow \texttt{westOf}(s, m).$$
$$\texttt{westOf}(s, foot) \land \texttt{hasFoot}(m, foot) \rightarrow \texttt{westOf}(s, m).$$
$$\texttt{southOf}(s, top) \land \texttt{hasTop}(m, top) \rightarrow \texttt{southOf}(s, m).$$
$$\texttt{southOf}(s, side) \land \texttt{hasSide}(m, side) \rightarrow \texttt{southOf}(s, m).$$
$$\texttt{southOf}(s, foot) \land \texttt{hasFoot}(m, foot) \rightarrow \texttt{southOf}(s, m).$$
$$\texttt{northOf}(s, top) \land \texttt{hasTop}(m, top) \rightarrow \texttt{northOf}(s, m).$$
$$\texttt{northOf}(s, side) \land \texttt{hasSide}(m, side) \rightarrow \texttt{northOf}(s, m).$$
$$\texttt{northOf}(s, foot) \land \texttt{hasFoot}(m, foot) \rightarrow \texttt{northOf}(s, m).$$

$$\texttt{locatedAt}(s, down) \land \texttt{hasDownstream}(r, down) \rightarrow \texttt{locatedAt}(s, r).$$
$$\texttt{locatedAt}(s, mid) \land \texttt{hasMidstream}(r, mid) \rightarrow \texttt{locatedAt}(s, r).$$
$$\texttt{locatedAt}(s, up) \land \texttt{hasUpstream}(r, up) \rightarrow \texttt{locatedAt}(s, r).$$

$$\texttt{eastOf}(a, x) \land \texttt{northOf}(x, b) \rightarrow \texttt{northEastOf}(a, b).$$
$$\texttt{eastOf}(a, x) \land \texttt{southOf}(x, b) \rightarrow \texttt{southEastOf}(a, b).$$
$$\texttt{westOf}(a, x) \land \texttt{northOf}(x, b) \rightarrow \texttt{northWestOf}(a, b).$$
$$\texttt{westOf}(a, x) \land \texttt{southOf}(x, b) \rightarrow \texttt{southWestOf}(a, b).$$
$$\texttt{northOf}(a, x) \land \texttt{eastOf}(x, b) \rightarrow \texttt{northEastOf}(a, b).$$
$$\texttt{northOf}(a, x) \land \texttt{westOf}(x, b) \rightarrow \texttt{northWestOf}(a, b).$$
$$\texttt{southOf}(a, x) \land \texttt{eastOf}(x, b) \rightarrow \texttt{southEastOf}(a, b).$$
$$\texttt{southOf}(a, x) \land \texttt{westOf}(x, b) \rightarrow \texttt{southWestOf}(a, b).$$

Fig. 4. Basic rules for geospatial reasoning about places

the definitions of River, Upstream, Midstream, and Downstream. This kind of intuition actually is formalized as a SWRL rule over a river r and its constituting parts up, mid, and $down$ as follows (assuming river r originates from two sources m and n):

$\text{River}(r) \wedge \text{hasUpstream}(r, up) \wedge \text{hasMidstream}(r, mid) \wedge \text{hasDownstream}(r, down) \wedge$

$\quad \text{originateFrom}(r, m) \wedge \text{originateFrom}(r, n) \wedge \text{endAt}(r, e) \wedge$

$\quad \text{hasPosition}(m, p) \wedge \text{hasPosition}(n, q) \wedge \text{hasPosition}(up, p) \wedge \text{hasPosition}(up, q) \wedge$

$\quad \text{hasPosition}(e, z) \wedge \text{hasPosition}(down, z) \wedge \text{meet}(mid, up) \wedge \text{meet}(mid, down)$

$\quad \rightarrow \text{Checked}(r).$

Similarly, for a mountain m, the part–whole constraints between the mountain top/side/foot and m itself are checked by the following SWRL rule:

$\quad \text{Mountain}(m) \wedge \text{hasTop}(m, t) \wedge \text{hasFoot}(m, f) \wedge \text{hasSide}(m, s) \wedge$

$\quad \text{hasPosition}(m, p) \wedge \text{hasPosition}(t, p) \wedge \text{hasPosition}(s, p) \wedge \text{hasPosition}(f, p)$

$\quad \rightarrow \text{Checked}(m).$

Note that in the above we use the concept Point to anchor the concept River: Individual points are used to relate a river to the river's constituting parts. So is the case for Mountain. We can relate rivers to mountains in a similar way, by connecting the place from which a river originates to certain mountains. This allows us to say that a river originates from these mountains.

Finally, we mention that some implicit facts can be derived directly by an OWL DL reasoner without the need of SWRL rules. For example, from the narratives:

> Kumuge Settlement is southeast of the river–cross of Mugua River and Cingshuei River. Kumuge is northwest of Tongment Settlement.

one can infer that Tongment is also southeast of the Mugua and Cingshuei river–cross. This is because the inverse property of northWestOf is southEastOf, and southEastOf is a transitive property. In addition, certain queries to a knowledge base can be formulated as class definitions (*i. e.*, new concepts with additional restrictions). By reasoning about whether the class is empty, an OWL DL reasoner will be able to deliver the results for us. For example, the query "what settlements are west of Sipan and are near by some rivers" can be answered by reasoning about the individuals in the following class:

> **Class**   WestOfSipanNearByRivers $\equiv$
> Settlement $\sqcap$ (westOf **has** Sipan) $\sqcap$ (bankOf **some** River).

We now return to the illustrative example we use in Section 2:

> Sikilian Settlement is located at the left bank of midstream Liwu River, and is about 1.6 kilometers to the northwest of the junction of its branch Wahei-er River. It is opposite to the small terrace slightly east of and below the Syuejiachang Station on Central Cross–Island Road. That is, it is at the mountain belly north of Mantou Mountain.

Geospatial knowledge in the above narrative is now formalized by the following assertions in our implementation:

$$Sikilian : Settlement.$$
$$Liwu, Wahei\text{-}er, \_U : River.$$
$$MidLiwu : Midstream.$$
$$Liwu\_Wahei\text{-}er : RiverCross.$$
$$Mantou : Mountain.$$
$$MantouSide : Mountainside.$$
$$Syuejiachang : Station.$$
$$\_T : Place.$$
$$\_P : Pair.$$

$$hasMidstream\ (Liwu, MidLiwu).$$
$$hasInflow\ (Liwu\_Wahei\text{-}er, Liwu).$$
$$hasInflow\ (Liwu\_Wahei\text{-}er, Wahei\text{-}er).$$
$$hasSide\ (Mantou, MantouSide).$$

$$leftBankOf\ (Sikilian, Liwu).$$
$$locatedAt\ (Sikilian, MidLiwu).$$
$$northWestOf\ (Sikilian, Liwu\_Wahei\text{-}er).$$
$$northOf\ (Sikilian, MantouSide).$$
$$eastOf\ (\_T, Syuejiachang).$$
$$consistOf\ (\_P, Sikilian).$$
$$consistOf\ (\_P, \_T).$$
$$inBetween\ (\_U, \_P).$$
$$sameAs^*\ (\_U, Wahei\text{-}er).$$

Note that in the above representation, we have ignored the distance (". . . about 1.6 kilometers to . . . "), some geographic features ("small terrace" and "Central Cross–Island Road"), and certain details about direction and orientation ("*slightly* east of" and "below"). In order to represent the narrative "it is opposite to the small terrace slightly east of and below the Syuejiachang Station", we use three auxiliary individuals: _T for the place of the small terrace, _P for the pair consisting of the place _T and the settlement Sikilian, and _U for the river that separates the pair _P. The last assertion states that river _U is the same as Wahei-er River. This fact is not in, nor can be inferred from, the given narrative (hence marked by a ∗). However, we can learn that Wahei-er River separates Mantou Mountain (where Sikilian Settlement is located) and Syuejiachang Station (where the small terrace is near by) by looking at a topographical map of this region. (See Figure 5 in Appendix A.) Therefore we assert _U and Wahei-er River are the same.

5   Related Works and Discussions

We now compare our approach to narrative geospatial knowledge to other works. First, the traditional GIS approach is to provide the humanity research communities with information systems to visualize spatial data, and to explore, analyze,

and visualize domain–specific knowledge in association with spatial data (often presented as maps). When dealing with domain–specific knowledge of a narrative nature, however, a GIS would need to quantify the narratives before they can be represented and analyzed. This can be a major hurdle.

There are works emphasizing on the summary and analysis of textual information, often in combination with gazetteers and other geospatial metadata, in the production of a geospatial web for the purpose of humanistic research. These works range from using RDBMS (Relational Database Management System) with RDF (Resource Description Framework), to constructing suitable spatial–temporal–object data model [15], to combining data mining techniques with location ontologies for the extraction of mutual relations among places in the cultural heritage domain [12], and to geovisual analytic approach to producing geo–historical context from implicit sources [19]. Approaches based on textual and metadata analyses for the extraction and representation of geographic information, nevertheless, may not rely upon or even require the use of well–developed domain ontologies. On the contrary, our work emphasizes the use of formal expressions for knowledge representation and reasoning, as well as the development of a domain ontology even though the data we are processing is of a narrative nature.

The approach presented here can also be compared to works that developed conceptual primitives and frameworks for the modeling of spatial–temporal activities expressed in natural languages [10,13,18]. However, here our narrative geospatial domain is of a much larger scale. In our case, settlements and their movements are expressed in geographic terms, while expressions on spatial–temporal activities tend to be framed in languages about personal or inter–personal space. Also related to our work are standard vocabularies such as GML [11] and SpaceML [7] for expressing qualitative and/or quantitative relationships among geospatial objects. These vocabularies, however, are for the expression of standard geometric, topological, directional, or even temporal relations which are of a technical nature. Our vocabularies are closer to the actual phrases used in narratives which are of a cultural nature.

Finally but not the least, there is a long tradition of conceptual modeling and ontological design for geospatial information, especially when associated with common–sense knowledge and the use of rule–based inference scheme [3,8,14]. Our works can be considered as a practical study in this direction, only that we use a standard (Web) ontology language and the associated rule language. Our ontology design is just at the beginning, and it has been developed by only a few in a top–down manner. It currently only handle certain geospatial relations. Issues of collaborative ontology development, geospatial object identity, entity naming and multilingual information processing, maintenance of consistency in the knowledge base, multiple–sourced knowledge acquisition, *etc.* are yet to be dealt with. These are critical issues to look into when building large–scale geospatial knowledge base.

References

1. OWL Web Ontology Language Semantics and Abstract Syntax (2004), http://www.w3.org/TR/2004/REC-owl-semantics-20040210
2. SWRL: A Semantic Web Rule Language Combining OWL and RuleML (2004), http://www.w3.org/Submission/SWRL
3. Cai, G.: Contextualization of geospatial database semantics for Human–GIS interaction. Geoinformatica 11(2), 217–237 (2007)
4. Clementini, E., Felice, P.D., Hernández, D.: Qualitative representation of positional information. Artificial Intelligence 95(2), 317–356 (1997)
5. Cohn, A.G., Bennett, B., Gooday, J., Gotts, N.M.: Qualitative spatial representation and reasoning with the region connection calculus. GeoInformatica 1(3), 275–316 (1997)
6. Cohn, A.G., Hazarika, S.M.: Qualitative spatial representation and reasoning: An overview. Fundamenta Informaticae 46(1–2), 1–29 (2001)
7. Cristani, M., Cohn, A.G.: SpaceML: A mark-up language for spatial knowledge. Journal of Visual Languages and Computing 13(1), 97–116 (2002)
8. Egenhofer, M.J., Mark, D.M.: Naive geography. In: Kuhn, W., Frank, A.U. (eds.) COSIT 1995. LNCS, vol. 988, pp. 1–15. Springer, Heidelberg (1995)
9. Freksa, C.: Using orientation information for qualitative spatial reasoning. In: Frank, A.U., Formentini, U., Campari, I. (eds.) GIS 1992. LNCS, vol. 639, pp. 162–178. Springer, Heidelberg (1992)
10. Hornsby, K.S., Li, N.: Conceptual framework for modeling dynamic paths from natural language expressions. Transactions in GIS 13(s1), 27–45 (2009)
11. Open Geospatial Consortium Inc. OpenGIS Geography Markup Language (GML) Encoding Standard (2007)
12. Kauppinen, T., Puputti, K., Paakkarinen, P., Kuittinen, H., Väätäinen, J., Hyvönen, E.: Learning and visualizing cultural heritage connections between places on the semantic web. In: Proceedings of the Workshop on Inductive Reasoning and Machine Learning on the Semantic Web (2009)
13. Klippel, A., MacEachren, A., Mitra, P., Turton, I., Jaiswal, A., Soon, K., Zhang, X.: Wayfinding Choremes 2.0 — Conceptual primitives as a basis for translating natural into formal language. In: Van de Weghe, N., Billen, R., Kuijpers, B., Bogaert, P. (eds.) International Workshop on Moving Objects From Natural to Formal Language, Park City, Utah, USA (2008)
14. Mäs, S.: Reasoning on spatial relations between entity classes. In: Cova, T.J., Miller, H.J., Beard, K., Frank, A.U., Goodchild, M.F. (eds.) GIScience 2008. LNCS, vol. 5266, pp. 234–248. Springer, Heidelberg (2008)
15. Sheth, A., Perry, M.: Traveling the Semantic Web through Space, Time, and Theme. IEEE Internet Computing, 81–86 (2008)
16. Liao, S.-c.: The migration and distribution of the East–Sedeq Atayal (I). Bulletin of the Institute of Ethnology, Academia Sinica (44), 61–206 (fall 1977)
17. Liao, S.-c.: The migration and distribution of the East–Sedeq Atayal (II). Bulletin of the Institute of Ethnology, Academia Sinica (45), 81–212 (Spring 1978)
18. Tenbrink, T.: Space, Time, and the Use of Language: An Investigation of Relationships. Mouton de Gruyter, Berlin (2007)
19. Tomaszewski, B.: Producing geo-historical context from implicit sources: A geovisual analytics approach. The Cartographic Journal 45(3), 165–181 (2008)

A   Sources and Maps

In this Appendix, we list the original paragraphs which we have used in this study as the sources of sample narratives on the locations of Taiwan aboriginal settlements. These paragraphs are taken from the comprehensive survey on the migration and distribution of the East–Sedeq Atayal conducted by Shou–cheng Liao (Masaw Mowna) in the 1970's [16,17]. The original paragraphs are written in Chinese; here we provide the English translations.

The nine paragraphs each describes the location of an aboriginal settlement. These nine settlements were identified by Liao from which people had migrated to Kele Settlement, a "multi–settlement community" in Hualian County at east-coast Taiwan [17, 146–147]. Liao described the locations of the nine settlements by the following:

1. "The land of Batakan ... is located at the left bank of midstream Liwu River, and is south of Sanjhuei Mountain. It is above the cliff on the riverbank opposite to now Jinheng Station on Central Cross–Island Road." [16, p. 121]
2. "Bolowan Settlement is located at the right bank of downstream Liwu River, and is north of Ta Mountain. The land is about 1 kilometer to the south of now Sipan Station on Central Cross–Island Road. It extends to the east to an unnamed river (called Puluowan River by the aborigines), and to the west to Yanzihkou." [16, p. 123]
3. "The land of Kalugi ... is located at the left bank of Liwu River, and is about 3 kilometers to the west of the junction of its branch Wahei-er River. It is slightly west of now Syuejiachang Station on Central Cross–Island Road." [16, p. 178]
4. "The land of Lausi ... is located at the left bank of midstream Liwu River, is around the area slightly northeast of the junction of its branch Laosi (called Pako by the aborigines) River. That is, it is east of and above now Heliou Station on Central Cross–Island Road." [16, p. 149]
5. "Lodox Settlement ... is located at the left bank of Liwu River, is about 7–8 kilometers to the north of the junction of its branch Tuosai River, and is southeast of Jiming Mountain. That is, it is at the place now called Shang-meiyuan." [16, p. 196]
6. "Mogolisi Settlement is located at the right bank of midstream Tuosai River which is a branch of Liwu River. It is southeast of Jhongyangjian Mountain, and belongs to the second terrace of now Shangmeiyuan (now renamed to Jhucunfennong). It extends to the south to an unnamed river (called Mokelisi River by the aborigines), and to the west to the mountain belly opposite to the junction of Tuosai River and its branch Dalama River." [16, p. 195]
7. "Sikilian Settlement is located at left bank of midstream Liwu River, and is about 1.6 kilometers to the northwest of the junction of its branch Wahei-er River. It is opposite to the small terrace slightly east of and below the Syuejiachang Station on Central Cross–Island Road. That is, it is at the mountain belly north of Mantou Mountain." [16, p. 178]
8. "Sowasal Settlement is located at the left bank of downstream Tuosai River, and is on the highland in between the river and the right bank of downstream

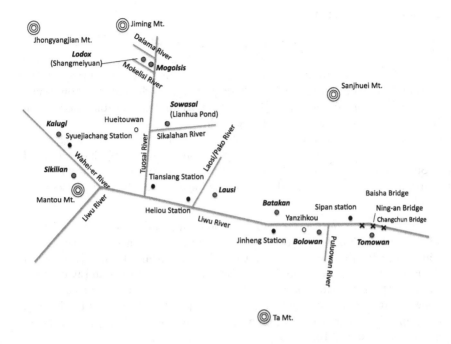

Fig. 5. Settlements are positioned according to geospatial narratives about them

Sikalahan River which is its branch. It is about 5 kilometers to the north-west of Tiansiang Station on Central Cross–Island Road, is above the river-bank opposite to Hueitouwan, and is at a place now called Lianhua Pond." [16, p. 155]

9. "Tomowan Settlement is located at the right bank of downstream Liwu River, and is northeast of Ta Mountain. It is at the mountain belly south of now Ning-an Bridge on Central Cross–Island Road. It extends to the east to Changchun Bridge, and to the west to Baisha Bridge." [16, p. 124]

We also provide in Figure 5 a topographical map of this region. The map is an abstract illustration. It shows the locations of the many rivers, mountains, bridges, stations, and other places that have been mentioned in the narratives and serve as the objects of reference when describing the the locations of the nine settlements. Results from this research are to be applied in situtation like this in which geospatial knowledge deduced from the narratives is used to position the locations of the settlements on the map. Figure 6 is a map taken directly from another part of Liao' ethnography [17, p. 135] on which Kanaongan Settlement is positioned at the northern river bank of the Dacingshuei estuary, in spite of the narrative in the same ethnography that it is located at the right bank of Dacingshuei River [17, p. 137]. See Section 1 for a discussion.

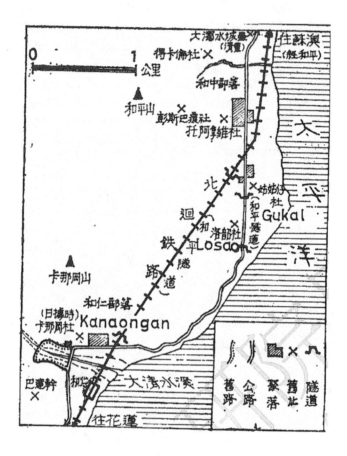

Fig. 6. An ethnographic map showing the location of Kanaongan Settlement